"I loved *A Loss for Words*. [The] style is brisk and clear and, it seems to me, never sentimental. Of cour⸺ ⸺e deaf, but as the book unfolds it⸺ ⸺s illuminated by [Walker's] style⸺ ⸺n who emerges to find her ow⸺ ⸺a character whom I admire as mu⸺  ⸺le

"In the end, I wanted to cheer Lou Ann Walker for having the gumption to write about a matter so close to her heart, learning to love and accept her parents as they are, not as she wished them to be. This is a gem of a book."   —*Glamour*

"Beautifully written and deeply affecting...There is humor in [Walker's] recollections but nothing lighthearted in accounts of crude or condescending reactions to her father and mother from indifferent people. Walker is candid in dealing with her own frustrations and the burdens of life with the deaf."
                                        —*Publishers Weekly*

"This warmly touching and humorous book is addressed to everyone who has ever lived in a family, which is to say all of us."
                                        —*Durham Morning Herald*

"So profoundly other is the unhearing culture...that moving it into a language we learn by hearing took both gifts and a nearly savage determination."   —*New York Times Book Review*

"Although I am a friend of Lou Ann Walker's, I did not dream there was such an affecting story to be told of a close, loving family in which both parents are deaf. I was thrilled to read her *A Loss for Words*, fortunate to find myself engrossed in a work so infused with tenderness, warmth, and humanity."   —Joseph Heller

A
LOSS FOR
WORDS

The Story of Deafness in a Family

Lou Ann Walker

PERENNIAL LIBRARY

Harper & Row, Publishers, New York
Grand Rapids, Philadelphia, St. Louis, San Francisco
London, Singapore, Sydney, Tokyo, Toronto

First PERENNIAL LIBRARY edition published 1987.

Designer: C. Linda Dingler

Library of Congress Cataloging-in-Publication Data

Walker, Lou Ann.
  A loss for words.

  "Perennial Library."
  1. Deaf—United States—Family relationships. I. Title.
HV2395.W34   1987   362.4'2'0973   85-42597
ISBN 0-06-091425-4 (pbk.)

    98 ✦/RRD H   20 19 18 17 16 15

For my mother and father,
for Kay,
and for Jan

Contents

In sign language, the gesture for "mother" is an open hand, the thumb resting gently on the cheek. The sign for "father" is the same hand, the thumb at the temple.

"Take care of the sense and the sounds will take care of themselves."

—Lewis Carroll, *Alice's Adventures in Wonderland*

A
LOSS FOR
WORDS

In order to protect the privacy of people who were part of my life, the names and certain individual traits of people in this book have been changed, with the exception of my family members and some prominent figures.

Prologue

I must have been about four the time Grandma Wells and I were cuddled up spoonwise in her bed. Grandma had her arms around me and the new ballerina doll I'd just received for Christmas. Grandma started humming, her voice, quiet and low, cracked. She breathed between words. Then her humming floated into a lullaby: "Sleep, my child, and peace attend thee, all through the night. Guardian angels . . . Soft the waking hours are creeping, hill and dale in slumber steeping. . . ."

Mom and Dad picked me up from Grandma's the next day. That night as Mom tucked me into bed, I asked her to sing me a lullaby, even though I knew she couldn't.

My parents are deaf. I can hear. And the fact of their deafness has made all the difference. It has altered the course of their lives, of my life, of their families' lives.

In a way we were outsiders, immigrants in a strange world. With my two younger sisters and parents, it was as if we were clinging together for safety. There were unbreakable bonds between us. Yet there was also an unbroachable chasm, for despite my parents' spirit and their ability to get along, their world is the deaf, their deaf culture, their deaf friends, and their sign language—it is something separate, something I can never really know, but that I am intimate with.

The best that can be said for deafness is that it's an invisible handicap. The worst, that it puts adults at the mercy of their hearing children, at the mercy of parents, at almost anyone's mercy. It is one of the cruelest and most deceptive of afflictions. It can emasculate men and devastate women. It is an impairment of communication. But it's not just the disfigurement of words and it's not just broken ears. It's most often a barrier between person and person.

I acted as interpreter and guide for my parents the entire time I was growing up. I was an adult before I was a child. I was quiet and obedient around people because I didn't know what was expected of me. Outside our house speaking and hearing seemed to be valued more than anything. And that's what we had nothing of at home.

I was the child who did all my parents' business transactions, nearly from the time I was a toddler. I spoke for my parents; I heard for my parents. I was painfully shy for myself, squirming away when the attention was focused on me, but when I was acting for my parents I was forthright. I made their doctors' appointments. I interpreted in sign language for my mother when she went to the doctor and told him where it hurt and when he told her what medicine to take. I told the shoe repairman what was wrong with a shoe. I told the store clerk when we needed a different size. It was me the garage mechanic would hang up on when I called about a transmission for my father. I was usually the one to relay to Mom and Dad that a friend had died when we received a call. I was the one who had to call up other friends or relatives to give them bad news.

A child doesn't know that his childhood is sad: it's just his life. I didn't realize that everyone didn't feel content at home and embarrassed and confused away from there. And it took me most of the first three decades of my life to figure out what those differences were between the deaf and hearing worlds.

When I showed my parents the first draft of this book, I was nervous. "Rough," I signed, one hand, clawlike, scraping the back of the other. It was a crude, ragged manuscript and I'd asked them to tell me exactly what they thought of it. It took them a couple of months to read it. Then they mailed me a letter and five notebook pages filled with single-spaced corrections, altering the sequence of events and reinstating the veracity to family legends other relatives had told.

They made only one substantive comment in the letter: "We hope no feelings hurt." They had just read four hundred pages in which their eldest daughter had ripped open their lives and held them up to scrutiny, in which I'd written every bit of the harshest truths about our relationships. I dredged up things I would never have mentioned in the ugliest of arguments. And their main concern was for other people.

At first my father was tentative about my writing this book. My mother was dead set against it. They are extremely private people. All the stares they get in public when they talk in sign has made them so. All their lives they've hidden away the insults and hurts other people caused them, and hidden the insults from themselves as well. They did not want to have their one bit of privacy made glaringly public. I did not blame them and I did not want to pressure them into allowing me to do this. Thinking it over, however, they changed their minds. "It may do some good for others. . . ."

This is a subjective work. In some places, having bottled up so much for so many years, I have vented my spleen. Writing it, I sat at my desk, endlessly typing the same pages over and over. It was as if, in the writing, the words had to come through my fingers for me to understand them, just as the meaning had to come through my fingers in signing. There were hours I cried, realizing how difficult the day must have been when my mother was taken away to school for the first time. I was crying from discovery—never once had my mother or father mentioned such things to me. Other times I was touched by the sweetness of a memory.

I discovered a great deal researching and thinking about this book: Nothing is as difficult as writing about your family, nothing can tear your soul apart quicker, nothing can make you cry and ache and wish you'd never begun. Nothing is as dismal as the feeling that you may be deceiving them, or even that you may be overpraising them and thus leaving them without the dignity they had before. Sometimes the people you've ignored growing up turn out to be more decent and solid than you could have realized. Often your childhood heroes melt away.

In some ways my parents' lives were exceedingly ordinary: plain stock from the Midwest, working-class people, ranch house, one dog, one cat, three children. And out of that comes the extraordinary.

Two ordinary people in extraordinary circumstances.

I can only hope that I've imparted something of what it is to be them.

I started out writing about my parents. I learned a tremendous amount about life. And I ended up finding out about myself.

Watching

1

Rearview Mirror

Mom and Dad drove me out to Harvard the fall I transferred. I'd never been east of Ohio. Looking back now, I know I was frightened. That day it came out as sullenness. I was scared of being a small fish in a big pond, terrified of being looked down on as the hayseed from Indiana. I was convinced that once the Harvard and Radcliffe administrations actually saw me, they would tell me to go home.

I was looking forward to getting away from home. Not from my parents. I was itching to break away from small-town thinking, from plainness, from flat land and houses that looked alike, from the constant interpreting, carrying out business transactions, acting as a go-between for my parents and a world that really didn't have much patience.

My head was filled with the aura, the stateliness of the Ivy League. Names resonated with import: Currier, Lowell, Winthrop. I could smell and hear things I'd never encountered, but in my imagination I knew they existed, and I felt sure that upon my arrival—if I wasn't sent home—wonderful happenings would occur. I wouldn't be burdened by timidity. No one would know of my mistakes unless I repeated them.

I'd just spent two years at Ball State University in Muncie, Indiana, with some vague idea that I wanted to be a teacher of the deaf. When the program turned out to be less than I expected, and when I didn't feel I was getting enough challenge in my other classes, I applied to four eastern colleges and was accepted. Harvard took very few transfers that year—the next year, none were admitted at all—and although the admission officers were very kind to me, all the literature they'd sent warned how difficult it was to switch colleges in midstream.

Now I looked up at the back of my parents' heads and I sank down low in the car's back seat. Filling out the application, I'd made prominent mention of the fact that they were deaf. The entrance essay, which was supposed to be about me, was actually about them. Many applicants use a father's or grandfather's degree to get them into the family alma mater, but neither of my parents had set foot in a college classroom. The irony that I was shamelessly using my deaf mother and my deaf father to get into Harvard was not lost on me. Neither was the fact that although I'd willingly and openly tell people they were deaf and I would briefly answer questions, I just wasn't going to say anything else. It was all too complicated.

Most of the sixteen-hour trip to Cambridge I brooded over a freshman reading list, the kind given out to high school seniors that includes all the books they should have read by the time they matriculate. I'd read very little of what was on that list. When I'd received it in the mail, I had gone to the library, taken out *Ulysses*, and despaired. I understood nothing.

I sat in the back seat for hundreds of miles, worrying that I'd have nothing to discuss at the dining table. And every once in a while I'd look up to watch my parents' conversations.

When the highway was deserted, Dad could comfortably shift his eyes from the road to Mom's hands. When traffic got heavy, he would have to watch the road and then his glances were shorter. If he wanted to pass a car, he'd hold up an index finger at Mom, signaling her to suspend the conversation for a moment. It was always easier for the driver to do the talking, although that meant his signs were shortened and somewhat less graceful. He would use the steering wheel as a base, the way he normally used his left hand; his right hand did all the moving.

Curled up in the seat, chin dug into my chest, I noticed there was a lull in the conversation. Dad was a confident driver, but Mom was smoking more than usual.

"Something happened? That gas station?" Mom signed to me.

"No, nothing," I lied.

"Are you sure?"

"Yes. Everything is fine." Dad and I had gone in to pay and get directions. The man behind the counter had looked up, seen me signing and grunted, "Huh, I didn't think mutes were allowed to have driver's licenses." Long ago I'd gotten used to hearing those kinds of comments. But I never could get used to the way they made me churn inside.

Mom was studying me. Having relied on her visual powers all her life, she knew when I was hiding something. "Are you afraid of going so far away from home? Why don't you stay in Indiana? This distance. Why wasn't college in Indiana good enough?"

"Mom. No! Cut it out."

She turned and faced front again, then she tried to distract both of us by pointing out a hex symbol on a barn.

Dad hadn't seen exactly what either of us said, but he'd caught the speed and force of my signs from the rearview mirror, and he could feel the tension coming from behind him. Mom had struck several nerves in me. Not only was I stepping into foreign territory—I hadn't been able to afford to visit any of the schools to which I'd applied—but also, back home in Indiana, none of my relatives or high school friends had been enthusiastic about my going east. To Hoosiers, Harvard means highbrow and snotty, too good for everyone else. Before I had left, Grandma Wells, my mother's mother, had admonished me, not once but several times, "not to get too big for my britches."

I couldn't concentrate on my reading or the view. Passing the rectangular green sign with the pilgrim hat welcoming us to Massachusetts, I felt the knot in my stomach grow. I was anxious to get there but dreading the moment of arrival.

When she saw the first sign for Boston on the turnpike, Mom tapped Dad's arm rapidly. I nervously flipped through my magazine. Soon we were crossing a bridge over the Charles River. Another sign said "Cambridge."

You have to see a lot of things to enjoy resting your eyes. You have to hear a lot of noise to appreciate the quiet. And you have to journey away from home before you can figure it out. Dad pointed to the sign and turned to look at me. In the corner of his eye was the beginning of a tear.

"I never thought my daughter would go to Harvard," he signed. "I'm so proud." To sign proud, he started with his thumbs at waist level and drew them up his chest, sitting a little straighter as his chest swelled up. Dad would have been content no matter what college or job I chose, but this was a dream he never had, that a family with no money, no connections, no education, could send a daughter to Harvard.

The strain was suddenly washed away. I smiled and leaned over the seat to kiss his cheek. Mom patted my hand and then my face. She was unaware of the long low hum of affection coming from her throat.

In Cambridge we were completely lost. I had a housing office address for my new dorm and an indecipherable map of the narrow, twisting one-way streets. Where I came from, things were laid out in a simple grid, and no one was ever on the sidewalks.

It turned out I was stuck in a left-over room in Jordan Co-op, and when we got there I didn't even want to look up at my mother's face. The place was pathetic, an unlived-in cinder-block dorm room with upended tables and chairs. There was litter in the hallway, the bathrooms were filthy, and the kitchen had an unidentifiable smell. My dreams of wood-paneled, leather-chair splendor were not coming true.

"Your room at Ball State was cleaner," Mom said, the sign for "clean" being one palm drawn neatly over the other.

I glared at her.

Only the overstudious "nerds"—whom Harvard half-affectionately termed "wonks"—had arrived this early. All the glamour that might still be attached to Harvardness disappeared when a peculiar-looking character with mechanical pencils in his pocket, frizzy hair, and enormous, twitchy eyes came into my room. He almost jumped as I began translating what he was saying into sign for Mom. I felt several worlds collide.

For both my parents, appearances are crucial. Most of the time the only impression they get is visual. They couldn't hear the accent in this fellow's voice. None of us had ever encountered anyone quite like him.

As soon as the last box and lamp were moved into the room, I told Mom and Dad they could go to their hotel. I told them I needed to plunge into Harvard life as soon as possible. But if I'd been honest with myself, I would have known that I was just as embarrassed having them meet the wonk as I was having the wonk and the rest of Harvard meet my deaf parents.

A couple of hours later we had an early dinner together and said our goodbyes. They were leaving very early the next morning for the drive back to Indianapolis. Standing on the sidewalk outside the restaurant, I was in a hurry to get on with things. We hugged and kissed, and Mom hugged and kissed me some more. She reached over to smooth the hair from my face.

"Promise you'll write often," she signed.

"I always do, Mom."

"You're coming home for Christmas, aren't you?"

"No."

"What?" Mom looked upset.

"I'm just kidding. Of course I am. How could I not come home and see you?"

Back in the dorm, I set about unpacking. I rearranged the daisies my sister Kay had gotten up at 3 A.M. that day to give me. As I set papers and supplies on my desk, a blue notepad fell open and for a second I was annoyed that my sisters had been doodling on the new pad a friend had given me as a going-away present. Then I read it. On the top sheet was scrawled: "Dear Sister, I'll miss you. Love, Jan."

My roommate had already arrived. Laura was the first person I'd ever met from California. A junior with long, wavy dark hair, she was beautiful. Her body was long and sinewy. She didn't look at all like my image of the studious 'Cliffie. We talked for a couple of minutes, and except for throwing a sheet over the top of her bed, she didn't bother to arrange any of her things. Instead, she went off to talk to the guy with the mechanical pencils for a while.

Laura came back to the room.

"Howard told me your parents are deaf and dumb. He said he saw you using sign language. That's pretty neat."

I cringed. I didn't want to sound priggish, correcting her for saying "dumb." "Um, well, I don't know if I'd exactly call it that. . . ."

At nine-fifteen, Laura announced she was going to bed and

took off all her clothes. Throwing them across the room onto a dusty chair she'd carried up from the basement, she dived under the sheet to rest on top of the bare mattress.

I'd never seen anyone sleep nude in my life. There was something in the nimbleness of her movement after she'd taken her clothes off that told me there were a number of things I was going to learn at Harvard that I hadn't foreseen. But mainly I was thinking about how prudish and midwestern it was of me to be so shocked at her unshaved armpits.

I put on my nightgown, dressing with my back to Laura, carefully pulling the gown to my ankles before reaching under to remove my skirt. I crawled into my crisp bed—made by my mother—and we talked for a few minutes more. Suddenly Laura kicked her foot up out of bed and hit the light switch.

In the dark I felt more forlorn than ever. I waited until I thought she was asleep, got up, put my clothes back on, and walked outside. I didn't know where I was going. I headed toward what looked to be the busiest street and discovered Massachusetts Avenue. Mom and Dad had told me that was where the Holiday Inn was located. I wandered up and down the sidewalk and found myself in front of their hotel.

I made my way to their room on the third floor and as I raised my knuckles, it dawned on me that knocking would do no good. I knew they were awake; I could hear the television. I took a notebook paper out of my purse and bent down to shove it underneath the door, working it in and out. There was no response. I tried crumpling up a small piece to throw into the room but I couldn't get it between the jamb and the door. I pounded on the gray metal, thinking they might feel the vibration. I must have stood there for twenty minutes, hoping Dad might come out to get ice from down the hall or perhaps go to the car to retrieve a bag. But he didn't. There was no way that night to get the attention of my deaf parents.

I walked back to my dorm.

Years later I would find out why the television was on that night in Cambridge. My father wanted to hide the noise of my mother's sobs from strangers. My mother cried for hours, my father trying in vain to comfort her. For both of them, this drive to a faraway school had awakened memories of their own trips to schools as children, memories they'd tried to keep hidden from themselves. Dad said the drive home was very long.

2

Missed Connections

My mother became deaf at the age of thirteen months when she had a relapse of spinal meningitis. Doris Jean was a wiry, precocious child who'd already learned to say "mama" when the illness struck. She was up walking and talking again after her first bout, and the doctor in Greencastle, Indiana, decided she was so improved she didn't need to have the complete series of injections. She lost her hearing during the second high fever. It was years before she said anything again and for the rest of her life few people would understand her when she did talk.

On an aberrantly cold March day in Montpelier, Indiana, when my father was two months old, his mother swaddled him in blankets and took him in her arms to her brother's burial in Odd Fellows Cemetery. George had died of pneumonia. My father, Gale, developed what they used to call the grippe. The fever burned out his auditory nerves, leaving him deaf before he was three months old.

Two chance happenings. Accidents. Both that doctor and my grandmother had the best of intentions. Yet those two decisions changed the course of my parents' lives and the lives of their families forever. My father's parents spent the rest of their days in self-recrimination. My father's father, H.T. (short for Harvey Trueman), was the Evangelical United Brethren minister in a northern Indiana circuit. Soon after he found out that his firstborn was deaf (my father's eldest brother, Garnel, was a blue baby and probably lost his hearing during childbirth), H.T. became a volatile and self-righteous preacher. Nellie May, my grandmother, having delivered seven children, two of whom were deaf, spent much of my father's infancy grieving, then threw herself into church work. Her lobbying on behalf of the Woman's Christian Temperance Union was particularly vociferous.

My mother's parents, on the other hand, seemed bewildered by her deafness. Grandpa Wells—Chester—started out as a farmer and ended up a foreman in a saw-manufacturing plant. I don't think I ever heard him use the word "deaf" all the time I was growing up. Instead, Grandpa looked to my grandmother—Ernestine—to take care of things. He felt she was the "smart" one—she'd had all A's in school and he had dropped out to go to work. She did the bookkeeping for the farm and the grocery store he ran. Later, she would work in the library at DePauw University in Greencastle. During the days when her daughter was young, Grandma would read books about Helen Keller and fret about how she could teach her own little girl. In old pictures I've found, Grandma had the beauty and mystery of a silent-film star, but in later photos, both she and my grandfather had grown self-conscious and the lines of their mouths drew tight. During our Sunday visits, Grandma would sometimes sit, leafing through old albums, unable to choke back her tears. Eyes downcast, she'd once again describe what it was like when her baby was sick. "I just don't know what else we could have done. We searched so hard. I wish there was a miracle. . . ."

* * *

Ten years after leaving for college, I was home again in Indianapolis, sitting in my parents' blue-carpeted living room, listening. I could hear the squeak of the furnace, the blower over the stove my mother had forgotten to turn off, the high-pitched electronic squeal of two table lamps, the overly loud, slightly irregular ticking of their kitchen clock. As deaf people, of course, they couldn't hear to silence the domestic sounds. Now that my sisters and I had moved out of the house, there was no one for the noisiness to disturb. To shut out the racket, I got up and turned off the blower and the lamps. Sitting back down in that darkened room, I realized that growing up in our house had been anything but quiet. It sometimes felt as if we were living in some kind of amusement park, lights flashing on and off the way they do around a carousel. Whenever the doorbell rang, a light in our central hall and another in the kitchen flashed on, and a loud buzzer went off for me and my sisters. (The man who'd installed our special doorbell was deaf himself and hadn't known how discordant the sound was.) When the phone rang, lamps in the living room and an upstairs bedroom went on and off. If the call was for my parents, we would hear a foreign bleep-bleep, which meant we had to hook up the TTY, a teletype-telephone device with a coupler for the receiver that translated the bleeps into typed-out messages. If Kay or Jan or I answered the phone in the hallway, we'd race back to the family room, where the three-foot-high converted teletype stood. We'd gotten the TTY when I was in high school. It was an early model—newer ones are portable, with electronic readouts—and whenever a call was going over the machine, the house had all the clatter of a newsroom. My parents even had a pulsing-light alarm clock, which they often silenced not by turning it off but by pulling the pillows over their heads. (Some of their friends who were heavier sleepers had vibrating-bed alarms, systems that literally shook them awake.)

Those weren't the only tricks we had. If my father was in the garage at his workbench, my mother would get his attention by switching the overhead bulb on and off several times in rapid succession. If my mother was reading the newspaper in the living room and he was sitting in his easy chair and wanted to talk with her, he would lightly stamp his foot on the floor until she looked up from her paper, at which point he'd wave his hand so she'd be certain which one of us was "calling" her.

Although my parents communicated solely in sign, even that language isn't completely silent. Signing is vivid, the hands often brushing against a shirt or thumping a chest. But if my parents were having an argument—easy to spot because of the velocity of their gestures—my sister and I would hear hands smacking hands. They could talk, but their words were unintelligible to anyone but my sisters and me. They used their voices simultaneously with their signs when they spoke to us. Otherwise, they talked only in emergencies or unusual situations. My mother has a voice that's soft and cooing as a mourning dove's, but she could nonetheless scold us roundly for coming home late. Her voice didn't carry, however. When she called out for us in the backyard, my name came out "looeen." "Kay Sue" and "Jan Lee," names specifically chosen for ease of pronunciation, turned into unearthly "Kehzoo" and "Zhanli." Often as not, if we were in another room, my father would come get us, yet if he did call, the timbre of his voice, slightly higher pitched than Mom's, reminded me of a harpsichord, one note played over and over again. I've often seen people react curiously when they hear my mother and father talk aloud (thus my parents usually relied on the more nearly certain paper-and-pencil method). Yet as long as I've been around deafness—around my parents, their deaf friends, my deaf aunt and uncle—every time someone refers to them as "mutes" or "deaf and dumb" or "the dummies," a violent, reactive surge of anger rushes through me. Those terms seem to be uttered so derogatorily; they seem to be clumping everyone together in one small, neglectable bin. Besides, virtually no deaf people have problems with their vocal cords; they just cannot hear to monitor their own voices.

As I sat in the living room, I realized that if all my parents had to endure—and all my sisters and I ever had to hear—was a little name-calling, life would have been much easier. On the face of it, deafness seems to be a simple affliction. If you can't hear, people assume you can make up for that lack by writing notes, that you can pass your spare time reading books, that you can converse by talking and reading lips. Unfortunately, things are always more complicated than they seem.

Until they're about the age of two, babies are tape recorders, taking in everything that is being said around them. The brain uses these recordings as the basis of language. If for any reason a baby is deprived of those years of language, he can never make up the loss. For those who become completely deaf— "profoundly" is the term audiologists apply—during infancy, using the basics of English becomes a task as difficult as building a house without benefit of drawings or experience in carpentry. Writing a grammatically correct sentence is a struggle. Reading a book is a Herculean effort.

Nor is lipreading the panacea hearing people would like to believe. "Bed," "bid," "bud," and "bad" all look identical on the lips because vowels are formed in the back of the mouth. The best lip-readers in the world actually "read" only twenty-five percent of what's said; the rest is contextual piecing together of ideas and expected constructions. The average deaf person understands far less. Thousands of times during my mother's life when she misinterpreted what someone said, she watched the other person grow impatient or become angry because she seemed so slow. She'd learned early on that it was easier—not only for herself but for everyone else as well—if she simply smiled and nodded. We would all pay a price for that later on.

Neither of my grandfathers ever learned a single sign. My grandmothers and some of my aunts and uncles got down the fundamentals, but none of them were expert enough to carry on a regular conversation. From time to time they, too, had to rely on notepads and pencils to get their points across to my parents. Worst of all, though, was my grandfather H.T. When he wanted something conveyed to my father, he usually told Nellie May. "Tell the boy to chop some firewood," he'd say, without even looking my father in the face, and then he'd turn on his heel and leave the room.

Helen Keller once said: "Blindness cuts people off from things; deafness cuts people off from people." Never having a real conversation with your parents or with your children is a good example. Yet some of the isolation is more complicated than that. People stare, even gawk at you in public when you're signing. And it's difficult to have an argument or stick up for yourself with a hearing person. Strangers inevitably feel awkward around you because they don't know how to talk to you. When people suddenly realize that you're deaf, they feel so embarrassed—for you, for themselves, for the situation. And so often you're just in the way. I've watched people in the street try to pass my father as he is walking. "Excuse me," they say, but of course my father doesn't hear that. When they finally get past, they turn and glare.

In a family where there is deafness, guilt is a constant undercurrent, tainting relationships, sometimes even shattering that family. My own grandparents constantly exhorted me to "be good," themselves feeling guilty for not doing more for their children, hoping somehow I would make up for things. They felt guilty for reasons they couldn't make clear to themselves. Time and again I heard my grandmother Wells say she would give her own hearing to make her daughter "whole." In her mind, deafness was some kind of divine retribution. My grandparents examined their lives over and over, wondering what it was that was forcing them to do eternal penance. The guilt feelings didn't do anyone any good. It breaks my heart sometimes when my father approaches me tentatively with a small request. "Excuse me," he signs, the curled fingers of his right hand lightly rubbing his left palm. "Could you please make a phone call for me? Would it bother you?" He feels guilty about needing help. And deep inside him, the guilt his own mother felt is still a burden to him.

It can't have been easy for my parents to have to rely on their three daughters to conduct all their business affairs. I signed before I talked, and from a very early age I ordered for Mom and Dad in restaurants and explained what they wanted to clerks. When I was about eight, they began giving me their letters so that I could correct the grammar. I don't think I ever minded doing anything they asked. (Like all children, I loved feeling important.) But in a few instances I was an unfaithful go-between. I could never bring myself to tell Mom and Dad about the garage mechanic who refused to serve them because they were deaf, or the kids at school who made obscene gestures, mocking our sign language. Not once did I convey the questions asked literally hundreds of times: "Does your father have a job?" "Are they allowed to drive?" Those questions carried an implicit insult to a family such as ours, which was proud and hard-working and self-sufficient. I reworded the questions if I had to interpret them. And I never allowed myself to think about the underlying meaning. What I admired about my father was his dignity, and my mother, her joy. I didn't want to upset the fragile balance. Children need to look up to their parents, even if it means doing some rearranging.

I once talked to a woman whose mother was blind. Her most poignant memory of childhood was a day she helped her mother, pushing her baby sister in a carriage, cross the street. She can still see the traffic heading toward them on the avenue as she focused all her strength into the arm guiding the carriage. She was struggling to remain calm. What was important was the appearance: making it look as if her mother was leading.

To the hearing world the deaf community must seem like a secret society. Indeed, deafness is a culture every bit as distinctive as any an anthropologist might study. First, there is the language, completely separate from English, with its own syntax, structure, and rigid grammatical rules. Second, although deaf people comprise a minority group that reflects the larger society, they have devised their own codes of behavior. For example, it's all right to drop in unannounced, because many people don't have the special TTY telephone hookups. How else could they contact their friends to let them know they're coming? If a deaf person has a job that needs to be done—from electrical wiring to accounting—he's expected to go to a deaf person first. The assumption is that deaf people won't take advantage of each other and that they need to support their own kind. The deaf world is a microcosm of hearing society. There are deaf social clubs, national magazines, local newspapers, fraternal organizations, insurance companies, athletic competitions, colleges, beauty pageants, theater groups, even deaf street gangs. The deaf world has its own heroes, and its own humor, some of which relies on visual puns made in sign language, and much of which is quite corny. Because deafness is a disability that cuts across all races and social backgrounds, the deaf world is incredibly heterogeneous. Still, deafness seems to take precedence over almost everything else in a person's life. A deaf person raised Catholic will more likely attend a Baptist deaf service than a hearing mass.

* * *

It's so easy for mothers and fathers to get along with their newborns, their tender, sweet-smelling babies. Then the world intrudes. And things get complicated very quickly. It was as if I'd been under anesthesia all the time I was growing up. I wasn't even aware that I was struggling to balance so many worlds. I performed my duties willingly but I had blinders on—which is just as well, because once I left home and started sifting through all that had happened, I began realizing how impossible things had been for Mom and Dad, how hard things still were and how hard they would always be. Every day they met with constant, irritating reminders of their short-comings, from the petty annoyance of not being able to ask for a cup of coffee in a restaurant, to the sobering knowledge that they couldn't hear cars careening around corners, and that deaf people had been shot in the back by policemen when they hadn't heard a command to halt.

Sitting in their neat, cozy house, Mom and Dad in their cocoon of sleep upstairs, I realized that what seemed as if it should have been different wasn't. And what seemed as if it should have been the same wasn't, either.

3

Mom

Indianapolis
September 1936

It will soon be fifty years since the September morning when my mother's parents drove her to the Indiana State School for the Deaf. She had just turned six. Grandma and Grandpa Wells were a young couple from the country who had never even met a deaf person before their daughter was stricken. And here they were leaving their only child, a bright, inquisitive girl, at a place they'd never seen.

The day before the trip, my mother watched as her mother packed all the new little dresses she'd been making for her over the summer, a smocked rose-print frock and a corduroy jumper among others. Her mother folded the clothes neatly into a large leather suitcase—a small overcoat, hat, and mittens at the bottom, pajamas and petticoats on top, shoes, covered in tissue, stuffed in the sides. She rolled up her daughter's dolls inside a favorite wool blanket, which she tucked into a paper bag, then put the suitcase and bag by the front door. The little girl was bewildered. She couldn't understand why her mother wasn't packing her own clothes. She sensed the sadness in the

24

house, but her parents had no way of telling her she was going away to school. And no way of letting her know she'd be able to come home again. They pointed at her, they pointed at the distance, and then Grandma cried. So the little girl, looking up at her mother crying, began to cry too.

The three of them got up early the next morning to make the two-hour drive to the school. My mother was dressed in a brown-checked pinafore. When the three of them arrived in Indianapolis, they met the superintendent of the school, and after a brief walk around the grounds, and a look inside some of the yellow-brown brick buildings, Grandpa got the suitcase and bag out of the trunk of the car and brought them into Mom's dormitory, laying them on a small black metal bed, one of a row of such beds. When he came back outside, my grandmother was tugging on the pinafore.

"Now, Doris Jean, you be a good girl," she was saying; then she stood and turned, telling my grandfather that if they didn't leave right away she would break down sobbing right then and there.

The superintendent, Dr. Raney, a large, kindly man, tried to distract the little girl, nudging her toward another girl so that my grandparents could slip out quietly. But Mom knew what was about to happen. As the car pulled away and the enormous iron gates shut behind, she threw herself at the black fence, shrieking, wondering why she'd been abandoned.

My mother had been thirteen and a half months old when she contracted spinal meningitis. It was mid-September 1931, and the Depression had just hit. She'd been sick a couple of days when a strange woman Grandma had never seen before came to the door of their house on Hannah Street. The small, dark woman asked for a glass of water and Grandma told her she mustn't come in the house because the baby had meningitis, which was highly contagious. Grandma led her around to the side of the house to a spigot and the woman pointed at the delicate white flowers growing in a border. "She's been eating those flowers, that 'Snow on the Mountain,' " the woman said. My grandmother said no, but the woman insisted. "I know. I'm a nurse," she said. Grandma, already fretful about her child, now grew angry. "No, it wasn't those flowers at all," she said. "If you're a nurse. you should know it's a sickness from a virus, not from any flower." The woman drank some water from her cupped hands, eyed my grandmother, then slowly walked away. The strange visitor, and what happened later, the relapse, have always been mysteriously intertwined in Grandma's head.

The doctor had sent all the way to Indianapolis to get medicine for Mom. When it arrived, my grandparents brought her in to the doctor's office immediately. Sweating, Grandpa held the baby's legs out, tight to the table, while the doctor injected the serum straight into her spine. They all knew that even a tiny jerk might leave her paralyzed or dead. Five times they went through that, the baby screaming from pain and fear. But a few days after the fifth shot, Mom was bright and alive and had started talking again. Grandma brought her in to the doctor's office. He looked her over, seemed pleased with himself, then threw away the rest of the medicine. Grandma glanced at the serum wrapper lying in the wastebasket. It read: "Caution: Five to seven doses." A week later the baby had a relapse.

Grandma can't remember exactly when it was that she found out her daughter was deaf. From the moment she read the wrapper, she had the feeling something would go wrong. It took a while before she would know what it was. The doctor confirmed her suspicions.

Grandma Wells went out and bought a book about Helen Keller and tried to teach her daughter to talk the way Helen had learned, holding the little girl's hand over her throat as she said words, then having the baby hold her small hand over her own throat.

"Oh, she was such a good girl," my grandmother would say, clasping her hands together. "She had temper tantrums, all right. Just like the ones I read about Helen Keller having. She broke pencils. She threw herself down on the floor. And she sucked her thumb. We had a hard time getting her to stop, and they had a hard time at the school too. But she was just so sweet. She wouldn't hurt a fly. And she was always so helpful. She'd see me sewing and she'd know whatever it was I needed and she'd run and get it.

"Why, one time when the old woman next door in Fillmore died—Doris Jean was about four; she didn't know what it was all about—I was standing in the kitchen over the sink and when I looked out, I saw that she'd picked flowers from our garden and she was in line, following the mourners into the house, her hands full of orange and yellow flowers."

The drive to the deaf school was too far for Grandma and Grandpa to go to see Mom more than a couple of weekends a semester. In a way it was a relief to my grandmother not to visit more often, the leaving was so hard. Mom spent fourteen years at that school. Each time she went home for the summer, she and her family, even her sister, Peggy June, eight years younger, would have to get reacquainted.

It was several years before my mother understood why she was at the school. At first she wondered if she'd been exiled for misbehaving or if something terrible was about to happen to her own mother and father.

Most of the houseparents were kindly, gentle people. But when Mom was six, her supervisor was Miss Pitman, a gaunt woman with long, dark hair tumbling nearly to her knees when she took out the pins at night. During the day Miss Pitman wore narrow, pointy shoes, and every morning she used to love braiding and curling the little girls' hair, especially Mom's, which was straight and coppery. There was one girl who always seemed to be in trouble—and she was so scared of the housemother that she would wet her bed. Mornings, Miss Pitman combed and yanked on her hair so hard that she cried, her hands to her head, twisting, trying to catch Miss Pitman's eye with a look that pleaded for release. When Mom recounts the story, her eyes fill with tears, her hands holding the sides of her head, reliving the pain. Once several of the girls told Dr. Raney how often Miss Pitman kicked them with those pointy shoes of hers. He didn't believe them. Until one day he saw for himself. Miss Pitman was fired immediately. By that time, however, she had terrified my mother out of her thumb-sucking habit.

Like virtually all schools for the deaf at that time, the Indiana State School emphasized oral skills: speaking and lipreading. In classes the children spent much of their time learning spoken language and written grammar and watching people gesture and point. They'd hold their hands over their teachers' throats, then over their own. And they wondered why the way they made their voice boxes bob was wrong and why the teachers' ways, which felt the same to their fingertips, were right. The teachers never seemed satisfied. Mom blew feathers over and over to get her breathing right for *f* and *p* sounds. Every year it was the same old flash cards, children pushing out dull-toned "buh-buh-buhs" whenever they saw a picture of a bumblebee, always the same drill, the same cards to learn the speech that came effortlessly to everyone else.

There is a controversy that rages in the deaf world and among educators for the deaf. It's a battle that can provoke me to fury. For centuries there have been two distinct attitudes about how deaf people should be taught: The oralists believe in speaking and lipreading without ever signing; and the manualists are pro-signing in American Sign Language (ASL). Both sides have their points. And there is also a camp of compromise—those who favor "total communication," or signing while speaking in full sentences. The oralists feel that unless deaf children master English skills, they will be outsiders all their lives. That point is valid, but to leave a child without language for a moment longer than is absolutely necessary seems cruel to me. There is also an underlying sense I get in the arguments of oralists that there is something terribly wrong with being deaf and wanting to hang around deaf people. The manualists, on the other hand, sometimes seem to be living their lives in a vacuum. They occasionally act as if they don't need contact with the hearing world. A few behave as if it doesn't matter that their English grammar is faulty or that they can't make themselves understood. And the problem with the middle-of-the-road approach is that combining the two extremes slows everything down. There have also been myriad systems for teaching language and phonetics to deaf children. None have been resounding successes. I objected to the rigidity of so many of these theorists and their systems.

At the Indiana State School for the Deaf, sign language was never a subject taught in class, and although it was used openly, it was considered something of a guilty secret, a crutch for those who couldn't master speech. Children taught each other and what they learned was reinforced by those few teachers who were deaf. Mom considered herself lucky. There were many schools, particularly oral ones which forbade signing, that didn't employ any deaf teachers. Some of those children grew up unaware of what happened to deaf people after high school. It seemed to them that all the other deaf kids disappeared forever at the age of eighteen.

There was no consistency among the educators. Many of the hearing teachers could barely finger-spell their names and they couldn't read the children's signs at all. It is easy to send out unintelligible pidgin signs and the children didn't dare correct the adults. Yet among hearing people, these same teachers often claimed they were adept at signing. At the Indiana school, if a child did poorly with a teacher who was an oralist, he or she was not moved to another class. More likely than not, the child simply repeated the entire year of school and hoped that he would more easily understand the next teacher.

Mom took to signing "like a duck to water," my grandmother said. Mom experienced moments every bit as thrilling as when Helen Keller spelled "w-a-t-e-r" under the pump. Where before the only way she'd been able to make contact with someone was to point or frown or smile, she suddenly found her world grew richer. She had people to talk to.

As the years at residential school passed, Mom enjoyed it more and more. There were only nineteen in her class—twelve boys and seven girls—and because the school rarely let children leave the grounds and parents didn't come to get them very often in those days, they all grew up nearly as brother and sister. It was true that every week and every year had a sameness to it. Wednesdays they ate meat loaf. For fourteen years. Friday evenings meant fish and chocolate pudding, and Mom would scoop some of that pudding into her milk so she could have both chocolate milk and dessert.

Mom and her friends grew into typical teenage girls. They gossiped about which teachers they liked and didn't like and who had gotten in trouble in what class. They had constant, all-consuming crushes on boys. In admitting a newly realized fancy for a boy, one girl would turn to another, her back blocking other people's views of her hands—the deaf version of whispering—and confide in small signs.

Mom was a gifted mime who loved doing her imitation of the heavy-lidded, slow-walking teacher for her girlfriends. She could break them all up into giggles with just a hint of her impersonation of the school secretary chatting on the phone, examining her nails, then looking bored and flouncing away when the boss came up to tell her to get back to work. Not being distracted by anything that was spoken made Mom's physical observations especially acute. And the actual signing styles of people she knew were also ripe for mime. Just as there are accents in speech, there are regional accents in sign. People from the South sign slower than people in the North—even people from northern and southern Indiana have different styles.

In addition to language and grammar, the children in junior high school and high school were taught some basic living skills. The girls ironed the boys' white shirts every week. The boys built wooden bookcases. The girls were also taught cooking and later on, in high school, many of them, including my mother, took a keypunch course to prepare them for outside jobs.

But during these years it was becoming increasingly hard for Doris Jean to come home for summers. She missed the small farm with her chickens and ducks and dogs and cats—and even a pet pig—that her family had when she was small. She has always had an implicit understanding of animals and they sense her gentleness. But her parents moved so often during her childhood that there weren't many children she knew in her neighborhood anymore. Very small children play side by side, so it doesn't matter that one is silent. Slightly older children figure things out intuitively. But as children age, talking replaces playacting. In one sense, Mom had been left behind. But in another sense, she had advanced in a way her family couldn't follow. In sign she could express complicated ideas and feelings, but none of that got across. She missed the company of the girls at school, the late-night secrets the girls signed in candlelight, even the classroom discussions, when all the students sat in a circle so they could see each other talk.

There is a story my Grandpa Wells used to tell about one of

Mom's vacations home. She was about fourteen the time she was helping him build a chicken coop.

"We were on the roof of that coop on the farm we had outside Greencastle," he said. "She was hammering nails and hit her thumb and I could hear her as clear as day. She yelled, 'Shit!' Well, I just don't know where Doris Jean could have learned that word. But she sure knew it that day." My grandfather laughed.

"I didn't say anything to her, you know," Grandpa said. "But I think she surprised herself."

Behind buildings in huddles and late at night, my mother and her friends had taught each other everything they could find out about the world, from romantic ideas about what happened on dates to even more romantic ideas about what life would be like after residential school. Curse words, though racy, were part of that magical outside world. Discreetly they practiced both signed and oral versions. The girls in Mom's class would remain friends for life.

In June of 1950 Mom was graduated from the Indiana State School for the Deaf. The class motto was "Rowing, not drifting." The class flower was the sweet pea. After a Chopin piano prelude was played for the parents in attendance, a teacher waved her hand, motioning for the class to march in. One of the students read a scripture from Matthew, then a poem was recited, "Step by Step," by a student with good speech. My mother signed the poem as it was read aloud and later signed the hymn "I Would Be True" in unison with the other girls in her senior class.

Suddenly, after years of regimented life, Doris Jean was completely on her own. Her parents wanted her to move back with them; she knew nothing about writing checks, looking for apartments, getting utilities hooked up, but there weren't any jobs for her in Fillmore, no deaf people, not much to do. She decided to strike out on her own. Soon she was living in an Indianapolis apartment with a deaf girlfriend who'd been a classmate. Mom was working as a keypunch operator for a company that made construction equipment. Both Mom and her roommate, Alice, were excited about the prospect of exploring the city. Even though they'd been in school all those years, they had rarely been allowed off the grounds.

They settled into their domestic routines quickly. After work they'd go together to do their grocery shopping. To find out how much she needed to pay, Mom would lean over and look at the total on the cash register. If the clerk said something to her and she didn't understand, Alice, an adept lip-reader, would sign it for Mom. They'd cook together, then head for a movie, one with lots of action so they'd understand the plot, or they'd take a walk, or perhaps just sit around and talk.

Their apartment didn't have a flashing-light doorbell and they didn't like leaving the door open because they couldn't hear someone entering. If one of them had a date for the evening, she had to get ready early in order to go downstairs and wait. If Alice was late in getting ready, she would ask Mom to go downstairs, let him in, and chat with the young man for a while. It was on one of those evenings that Alice's temporary stand-in struck the fancy of her date for the evening—my father.

4

Dad

Northern Indiana

It was 1927. Everywhere else people called it the Roaring Twenties. In Montpelier, Indiana, which had once boasted a fancy gilded opera hall, the Depression had hit two decades before, when all the oil wells dried up. The town itself, thanks in part to my grandmother Nellie's WCTU efforts, was also dry. In addition to his circuit ministry, Grandpa Walker was trying to make a go of the mortuary business he'd started fifteen years before, which had never once gotten out of the red. He also latched onto frame-making and farming, and just about any money-making scheme he could think of to feed the six children and the new baby.

Grandma Nellie was short and round and lively. Her trademark was to burst into a house calling: "Yoo hoo!" She was the backbone of the family, bustling about, raising the seven children, tending her garden, helping out with the business in every way she could. More often than not, that meant laundering bloodied linens for the Walker ambulance service.

Grandpa Walker, on the other hand, was tall, with a long, stern face. In his later years when he was stooped and his white

hair thin, I'd see H.T. pinch my sister Kay's shoulder blades—
she was only four then—and, with his tongue caught between
his front teeth, squeeze until she shrieked with pain. "Oh, that
couldn't have hurt at all," he'd mutter, letting her go so he
could fuss with his pipe. H.T. was a believer in education and
hard work, although as my aunt Gathel would put it delicately,
"He was a great one for saying one thing and doing whatever
he pleased." Nevertheless, H.T. was proud and he instilled in
each of his children the idea that it was an honor to be a
Walker.

Whether it was pride or peculiarity, however, that spurred
him on when he gave them their Christian names, no one will
ever know. His firstborn he named Garnel Boyd, and in quick
succession came Ghlee Delight, Garl Dwight, and Golden
William. With his fifth, Nellie insisted on having some part in
the process. He let her have the middle names: Gathel May,
Gene Iris, and my father, Gale Freeman, Freeman being Nellie's maiden name.

The first four, in addition to their distinctive names, also had
unusual name signs, the special gestures used to abbreviate a
person's name when signing. Garnel, my father's eldest
brother, by bizarre coincidence, was also deaf. Garnel's name
sign was two fingers tapping the side of the forehead. Ghlee's
was a "g" handshape—the gesture people use to indicate "a
little bit"—held at the chin because hers was pronounced.
Garl's sign was a hand held sideways at the middle of the forehead, a sign denoting "great intelligence." Golden, who was always called Bill, was referred to with another "g" handshape,
this time turning up the end of the nose. The sign was both a
pun on his nickname, and a reference to the actual shape of his
nose, the most upturned of the characteristic Walker ski slope,
which my father also had until a too-vigorous high school
football tackle.

* * *

There's a tremendous amount of folklore surrounding deafness. From ancient times through the early part of this century, the mysteriousness of the condition made the public view deaf people as either prophets or devils. James Joyce once said, while passing in front of a skeleton, that he was superstitious about only one thing—"deaf mutes."

Often as not, the actual cause of deafness is presumed rather than pinpointed. Over the centuries people have blamed everything from eating a green chestnut to being cursed by a gypsy. Others have cited being frightened by a burglar, consuming improper combinations of food, and thinking and speaking impure thoughts as direct causes of deafness. My grandmother believed that the source of Garnel's deafness was the few seconds at his birth in 1908 when the oxygen supply to his brain was cut off. There's no way of verifying that. My father, she insisted, was not born deaf. "Insisted" because there was an even greater stigma attached to genetic deafness (and the family) than there was to acquired deafness. As far as she and my grandfather could remember, there had never been any deafness in either of their families.

Garnel was about two years old when Nellie realized he was deaf. She was distraught until one day, she said, the Lord spoke to her. He told her that if she did not accept Garnel as he was, and do for him and make the best of it, He would take the baby away from her.

Grandma Nellie never mentioned whether God spoke to her again, not even when my father's deafness was discovered. Although by that time the family had had nearly twenty years to adjust to deafness, in some ways my father's must have been the harder to accept. First, Nellie must have shouldered some of the blame for having taken that tiny baby out in the bitter cold, and then watching him sicken. What was more difficult, though, was that the family knew what to expect: trouble. Garnel was a difficult child. He threw violent temper tantrums that continued well into adulthood. In short, Garnel was exasperating. And the Walker family was slightly nervous about the likelihood of having to contend with another trying member. It's difficult to know where personality leaves off and circumstance begins, but the brothers and sisters still marvel at how different Garnel and my father turned out.

When Dad was born, he was a fat baby. At first the family called him Butterball, soon altered it to Puffball—a variety of local mushroom—and then shortened that to Puff. Even Nellie, who objected to nicknames, seemed to like this one. There was something right about it.

It took them a while to discover Dad's deafness. The signals were confusing. His cries sounded like any baby's. Crying is a reflex all babies have, even if they can't hear. Speech came to Dad slowly, but he did make some noises, and a deaf baby's voice isn't all that different from that of a hearing baby. If someone walked into the room behind him, he usually didn't respond. But of course he did if he saw the person. He was alert and active. He'd already begun compensating for his deafness.

Eventually the realization that something was wrong was unavoidable, and his parents took him to a doctor, and then another and another. The answer was always the same: The auditory nerves were destroyed. There was no cure. They'd have to live with it.

In some ways Puff had a typical Indiana boyhood. On the family farm, he was a hard and willing worker. He helped milk the cows, put up the hay in the barn, pump the water for the animals and for washing clothes outside. He'd go with his sisters Gene and Gathel to collect the eggs and feed the chickens, hogs, and other animals. If he was home when the weather was cold, he'd help with the butchering and soapmaking.

Puff, lanky, towheaded, and athletic, was often content to

play by himself, but the difference between him and other little boys was that he had to pay far more attention to whatever it was he was doing. He was always watching to see what was happening so he could join in. And he learned to be vigilant because he couldn't hear warnings.

Once, when he was eleven, he dashed off the porch at home in Montpelier as a car came careering around the corner. His mother stood screaming but helpless. Puff was fortunate; he bounced off the side of the car. If he'd run out a second sooner or if the car hadn't swerved, he would probably have been dead. It was a valuable lesson. He willed himself to be even more alert and agile from then on.

Puff was about five when Nellie insisted they take him to the Mayo Clinic in Rochester, Minnesota. He was having slight breathing problems and the family wanted his hearing checked again. H.T., Nellie May, and Puff were gone for five days. A battery of tall men in white coats spent hours poking and prodding Dad's ears, throat, and chest. Nellie had left full of hope, and when she returned she was despondent. Dad remembers her trying not to show her disappointment.

When Garnel was at the state school for the deaf, H.T. had a falling-out with the superintendent, and so after returning from Mayo, H.T. announced he would not allow Puff to attend. H.T.'s grudge would have cruel implications for my father. Dad's days at Huntington Street public school in Montpelier were a complete blur. He sat in class trying hard to pay attention and figure out what was going on. The teacher faced the blackboard so often when she talked that he couldn't lip-read her. Because his last name started with a *W*, he was in the last seat in the last row, so he could hardly see her anyway. His grades were uninspiring C's. Second grade was worse. His brothers and sisters pitied him for getting stuck with the sour-faced Miss Trent.

Luckily, by the time Dad was ready for the third grade, a new general superintendent of the deaf school was appointed, the same Dr. Raney who a couple of years later would be showing my mother's parents through the imposing brick buildings. Whereas my mother was terrified of the place, Dad, with his mischievous, gap-toothed grin, felt tremendously relieved the moment he was enrolled. The time at Huntington Street proved to have been a waste. Dad had to begin all over, a tall eight-year-old among the first-graders. He'd had the advantage of having a brother who was deaf, which meant some of the people in the family had been able to communicate with him when he was quite young. After his first year, he was promoted quickly. The school wasn't at all rigorous, but the children were well disciplined and worked hard.

Dad flourished at school. The mahogany bookshelf he made there with its intricate cutouts still stands in our house. His class was small—only nine people—so the students had plenty of individual attention. Sometimes too much. It was a cloistered life. To leave the grounds, children had to be accompanied by a relative because officials worried about what might happen to them out on their own.

One weekday in the middle of the term, H.T. showed up at the deaf school and had Dad pulled from his class. Walking down the long, dark hall, Dad looked past his father through the doorway to see his aunt Emma sitting in the car, her face in her hands, her shoulders shaking. There was a man he didn't know sitting in the back seat. H.T. pointed to the place next to the man, then he got behind the steering wheel. They drove for hours. Dad recognized the road. They were on the way to Emma's farm near Amboy, Indiana, way in the northern part of the state. Puff still didn't know what was wrong. Emma just cried and twisted her handkerchief, and when she talked, her mouth was contorted by the sobs. The strange man seated next to the boy just stared out the window.

Arriving at Emma's farm, Dad figured it out. Some men had removed the picture window from Emma's house and now they were trying to fit through a giant coffin. A neighbor took Dad out to the field next to the ten-foot-tall corn picker. In pantomime he showed how Uncle Hen—Henry, really—an enormous man who used to keep skinny little dogs, had been standing atop the picker as it moved down the cornfield. A sharpened metal tooth caught the edge of Hen's coat and yanked him into the machine's jaws.

By high school, Dad had obtained permission to take the interurban tram by himself to go home for special visits. Once in a while his brother Bill would stop by to see him if he was in town, but it was Gathel, in nurse's training at Methodist Hospital, who came most often, and who took Dad out for ice cream sodas.

Senior year of training, Aunt Gathel was in surgery, working with Dr. Sage, one of the best ear, nose, and throat men in the state of Indiana. She had talked to him at length about Puff's deafness and asked if he would take a look at him. Gathel asked Nellie May and H.T. to make the appointment and come down from Montpelier to take Puff to the doctor. The memory of the trip to Mayo still seemed fresh and they didn't want to go through it all over again. Gathel pestered them until they relented.

Later, Gathel found out that Dr. Sage said there was no help for Puff's deafness. "I felt so bad," she admitted years after. "I felt that I had built up Puff and the folks—especially Mother—for another letdown. I felt very guilty and very sad about that."

Oddly enough, Dad didn't feel at all sad. He had never let

himself get his hopes up, and so he wasn't at all disappointed then or any other time. He only felt bad because Gathel felt so guilty.

Dad hardly gave the doctor's appointment a second thought. He was too excited about graduating from high school. He'd learned a trade to prepare himself for going out into the world. He was taught how to operate a linotype machine, an enormous contraption that converted molten lead into type slugs. The slugs were then used in the newspaper printing process. The job was considered ideal for deaf people because they were careful, patient workers who wouldn't make too many typos, and presumably they weren't disturbed by the noise of the heavy machinery. Dad was looking forward to being out on his own, away from school, earning a salary.

There would, however, be one more incident with doctors, soon after. Dad found this one particularly amusing.

At the very end of the war, Dad was called before the Blackford County Selective Service Board. Most of the men had known the Walker family all their lives, but the board was desperate for recruits. At one point Dad was writing a note to an officer when someone behind him slapped a hand down on the table. My father started and whirled around. Like most deaf people, Dad is sensitive to vibrations, not because his ability to feel them is heightened but because he pays more attention to the sensations he receives. Also, although he describes himself as "stone deaf," there is virtually no one who has an absolutely flat audiogram. Somewhere there is some tiny perception of something akin to noise in everyone. For Dad, it happens to be a very loud bang, just like the one on the draft board's table. When another officer banged, Dad jumped again. The board required medical reports no matter how well its members knew the young man, but because their suspicions were aroused about the severity of Dad's deafness, they forced him to see several doctors. Finally—according to family legend—it was not deafness but his very flat feet that kept Puff from being drafted.

5

Good Sound

Mom was nineteen, nearly twenty, when she finished high school. Dad had been working as a printer at the Bluffton *News-Banner* for three years when he went to pick up Alice for their date in her apartment off Central Avenue in Indianapolis. It was the fall of 1950 and Alice made the mistake of leaving Doris Jean and Puff alone. As my mother signs, blushing, "It was love at first sight," the first two fingers on each hand facing each other, a little surprised, at "sight." Every weekend Dad made the two-and-a-half-hour drive from Montpelier to Indianapolis in his green '47 Studebaker. All that winter they went out for dinner dates, and in May of 1951, Dad asked Mom to go on a picnic to Qualifications Day at the Indianapolis 500-Mile Race. Dad knew it was going to be a special day. He brought along his camera and asked several people to take pictures of them. Mom wore a crisp white blouse with tiny capped sleeves at her shoulders and a long, dark, flowing skirt. Dad rolled up the sleeves on his white shirt. He had on good pleated trousers.

In the middle of the infield, they spread a blanket and set

down a red metal picnic hamper, Mom happily pulling out sandwiches and salads she and Alice had made. A few feet away, cars were roaring by, grinding gears, smashing into cement walls. Dad gently touched Doris Jean's arm while she was looking for the mayonnaise. Dad asked Mom to marry him. Actually what he signed was: "Can we marry soon?" The sign for "marry" is two hands circling then firmly clasping each other.

The crowds were all around them, people playing catch, shouting to each other above the din of the track. Yet it was an intimate conversation because no one around them could understand what they were saying to each other. Mom shook her closed hand up and down—"Yes."

The next weekend the two of them drove to see her parents. Doris Jean went up to her mother, pointed at her boyfriend, pointed at herself, and, using the manual alphabet, slowly finger-spelled "m-a-r-r-y." Grandma Wells clapped her hands and broke into a grin.

"Chet," she yelled at my grandfather, "they're getting married!" Immediately he went over to my father and pumped his hand. Pursing his lips, he said "Good" and nodded his head. My father, describing the meeting later, would say: "They were appreciated we asked them for the marriage."

Doris Jean and Puff set the date for a respectable but not overly long four months thence. The wedding was to be held in Fillmore, just a bulge on the highway, where Grandma and Grandpa Wells were living. H.T. was to officiate. Because they couldn't use the phone, Mom and Dad made several trips to Fillmore and Montpelier to check on flowers, invitations, and details for the reception.

The week before the wedding, H.T. typed up their vows so Mom and Dad would know exactly what to expect. It was a traditional text. H.T. began by asking: "If anyone has any objection, let him or her speak now or forever hold their peace. . . ." The ceremony ended with the Lord's Prayer. When Dad showed the vows to a friend, he excitedly wrote across the bottom of the single-spaced page: "Is this good sound? I think it is. Oh Boy! We'll be husband & wife."

They were married in the tiny white frame Methodist church on Main Street. My mother wore a virginal white lace gown with pearl buttons. She was a shy bride. Dad wore a gray suit and wide tie. Grandpa Wells gave the bride away. It was "simple and nice," my Grandma Wells once told me. "Real pretty. Nothing fancy." She and Grandma Nellie both cried.

The attendants, both deaf, were two of Mom's and Dad's best friends. There was also an interpreter, Miss Christian, an older woman who wore gloves and a big hat festooned with birds that bounced as she signed.

H.T. could starve you to death while he was saying the blessing before a meal, and I imagine that at the wedding of his youngest son, he got pretty wound up. But Mom and Dad were so excited, they could hardly focus their eyes on the interpreter's hands. Like every young bride and groom, they barely knew what the preacher had said.

Prompted at the right moment by Miss Christian, they each signed and uttered a muted "I do." After the ceremony, everyone, friends and relatives, went back to the Wells house to eat a ham and potato salad lunch, followed by cake and punch and mints that had been made by my grandmother. She'd also had to clear most of the furniture out of one bedroom to hold all the presents that had been brought. Mom and Dad fed each other bites of wedding cake. The pictures show Dad cupping his hand to his chin as Mom coyly overstuffed his mouth.

Mom changed into a gray suit she'd bought for her trousseau and then the two of them got into Dad's almost-new black '50 Chevy. As Dad turned on the ignition, there was an explosion under the hood. Mom, easily frightened, jumped out and clear of the car. Dad laughed. He knew someone had lit a pop-ping device. He coaxed her back into the car and they took off, waving. But as they did, everyone at the reception began laughing. One of my father's brothers had put rocks in the hubcaps. Only the plan had been foiled: Mom and Dad didn't feel the rocks banging—and they certainly didn't hear them.

My aunt and uncle and a couple of their friends, thinking the rocks might dent the hubcaps, hopped into a car to stop Mom and Dad. Dad drove faster. He was sure they were about to pull another trick. The race continued through the back-country roads, Dad having no intention of letting them smear him or his car with shaving cream or whatever. Finally, my aunt pulled up alongside the Chevy, pointed toward the hub-caps, frowning and frantically shaking her head "no." Dad pulled over. Sheepishly, the fellows in my aunt's car removed the hubcaps, derocked the car, and Mom and Dad took off, headed south for the Great Smoky Mountains.

Dad loves driving, especially through winding roads and mountain passes. The scenery was beautiful. In the trunk were the heavy brown leather suitcases they'd bought together. Mom had made a neat list of each item she'd purchased: stockings, hat, girdle, gloves, and next to the items were the prices she'd paid from her salary. Everything for their life was to be shiny and new, down to their underwear.

The two of them were delighted to be free, away from the regimentation of school and their parents' watchful eyes. Except for a few trips to the hospital, for the next thirty-odd years, they were rarely apart overnight.

On the second day of their honeymoon, they stopped at a mountaintop gift shop, looked around, but didn't buy any-thing. As they were leaving, though, the owner rushed up and grabbed my mother, startling her. My mother looked down. The shopkeeper was pressing something into her hands, a pair of tiny moccasins to fit a baby. My mother looked up at her questioningly. The woman started speaking, but Mom smiled, shook her head, and pointed at her ear. The woman picked up my mother's left hand and tapped her shiny rings, pointed at my father, and then shook their hands. It was a wedding present.

For the first time in their lives, Doris Jean and Puff had someone to talk to twenty-four hours a day, every day of the year. They no longer had to worry about teachers slapping their hands for using them to talk. The curtain that language had drawn between Mom and her family, and Dad and his, was no longer there. The inability to hear is a nuisance; the inability to communicate is the tragedy.

American Sign Language—ASL—is a language unto itself, with its own syntax and grammar. Adjectives follow nouns, as in Romance languages. In sign, one says "house blue," establishing a picture of what is being described and then embellishing on that. Many sign language "sentences" begin with a time element and then proceed with what happened, thereby conjugating the verbs. The movement of the shoulders, the speed of the hands, the facial expression, the number of repetitions of a sign, combine with the actual signs to give meaning to the language. Signing is precise. The casual gestures hearing people make when talking have no meaning in sign language. Hearing people who do learn sign usually practice "signed English," a word-for-word coding of English into signs, but that translation sorely limits the language. In some hands, signing is an art equal to an actor's rendering of Shakespeare. It is not just swoops and swirls but an enormous variety of expression, just as a great actor's delivery is completely different from some ham's idea of haughty speeches.

The creativity can be remarkable. A person can sculpt exactly what he's saying. To sign "flower growing," you delicately place the fingertips at each side of the nose as if sniffing the flower, then you push the fingertips of one hand up through the thumb and first finger of the other. The flower can bloom fast and fade, or, with several quick bursts, it can be a whole field of daffodils. In spoken English, most people would seem silly if they talked as poetically as some supposedly illiterate deaf people sign. With one handshape—the thumb and little finger stretched out, the first finger pointing forward— you can make a plane take off, encounter engine trouble and turbulence, circle an airport, then come in for a bumpy landing. That entire signed sentence takes a fraction of the time that saying it out loud would.

The face and body convey nearly as much as the actual signs. A raised eyebrow can completely alter the meaning of a sentence. Recounting a conversation, the signer shifts his upper torso just slightly, thereby doing away with the need for unwieldy "he said . . . then she said" constructions. People stutter in sign. There are even sign language equivalents for spoken sentence fillers, such as the irritating "you know," "well," and "I mean." Instead of using those phrases, the signer repeats a particular gesture, such as the hand flipping over, palm up. And just as surely as the British can pinpoint a person's station in life and place of birth upon hearing a couple of sentences, a signer can do the same upon seeing a couple of phrases. Signing can be small and intimate or big and brassy. It can convey every nuance imaginable. The rules for inventing new signs are strict. Mom and Dad, of course, had signs they invented for special things—the name of their street, for example. And they made up private names for each other.

Mom and Dad moved into a rented house in Montpelier and began saving for their own home. They busied themselves setting up housekeeping, and within six months of their marriage Mom became pregnant. She was nearly eight months pregnant when one night my father had a sudden attack. Mom ran to get help—of course they didn't have a phone. It was appendicitis.

Dad was rushed to the hospital for an appendectomy, which went well. Mom had been so frantic she'd left her keys inside the house. (Dad, as always, had so much presence of mind he'd locked the door even when he was doubled over with pain.) She had no way of getting inside, and her car keys were on the same ring. My mother never wanted to cause any trouble and wouldn't dream of bothering anyone. She didn't want people to think she was less capable or more dependent because of her deafness. So she broke the window and crawled in, delicately maneuvering her girth over the jagged shards of glass.

6

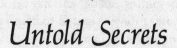

Untold Secrets

Hartford City, Indiana
December 1952

As my mother gave the final push that landed me into the
world, Dr. Douglas, her obstetrician, stood up, looked at her,
and traced the hourglass figure a man makes when he sees a
curvaceous woman. That was his way of telling her she'd had a
baby girl. The nurse took me off to clean me up, and Dr.
Douglas went over and clapped his hands near my head to see
if I'd respond, a crude test which probably didn't prove anything, but then he walked back over to my mother, smiling, as
he pointed to his ears and nodded. Everyone in the delivery
room was relieved I was a "normal" baby.

The notion that the womb is a silent place is pure fantasy. If
a person dives under water, he hears very little because sound
is muffled by the cushion of air remaining outside the eardrum.
A fetus has no air bubble outside its ear, and water conducts
sound better than air. Thus, from four months on, the unborn
baby can probably hear any number of things: its mother's
voice and the rumblings of her stomach, music, traffic, conversation, radios, even the cries of other children. In fact, medical

researchers have determined that the fetus hears before it sees: the eyelids don't open until about the seventh month, and at birth, the sense of hearing is better developed than that of vision. The newborn's eyes are open, but when the baby responds to something by turning its head, it is actually searching for the source of sounds. Babies prefer voices over other noises, and women's voices over men's. Armed with that knowledge, it's little wonder that some infant psychologists speculate that much of our personality may be determined *in utero*. In the course of my own idle speculation, I've wondered about the impact of extended periods of silence on my sisters and me, whether we could differentiate my parents' muted voices from other voices as soon as we were born, whether we had some innate appreciation for what things would be like for us.

Dad passed out cigars to the men at work and cradled his arms to let them know his wife had had their baby. He pulled out the small white notepad he always carried and wrote: "girl." Underneath, he wrote: "Lou Ann."

My grandmother Wells had helped Mom and Dad choose that name. They'd needed words they could pronounce easily, a name with few syllables and one that didn't have any *t*'s or *r*'s or *p*'s.

I'm told my mother just radiated happiness my entire infancy—except when the hint of red hair wore off and she had to tape a bow to the top of my head to clue people in on my gender. I'd ended up with my mother's Irish coloring and my father's features—a roundish face, turned-up nose, and dimpled chin—and everyone was happy about the compromise.

Before I was born, Mom and Dad paid sixty-five dollars, more than a week's salary, for a baby cry box, specially designed to alert deaf parents when their children cried. They placed the dark-brown plastic box, shaped like a radio, next to my crib and wired it to a lamp by their bed. As I cried, the box transmitted an impulse to the bulb, which flashed on until I paused for breath. Every night Mom and Dad were privy to their own light show. The device was so sensitive to the noise that as I sputtered, the light sputtered. My mother hadn't read Dr. Spock, probably hadn't even heard of him, but her ability as a mother was instinctive, loving and patient, and she usually carried me from room to room as she cleaned house so she could keep an eye on me. If I was napping, she frequently came in to check. And those times at night when I cried for no good reason, when I'd been fed and diapered and comforted and there was nothing left to do but wait out my tears, the flashing light strobed through the darkened house. Mom and Dad had one advantage over most parents, however. They didn't have to listen to my wails. They could pull the plug if they wanted, although I doubt they ever did.

When I look at the old pictures of them, pushing me in a stroller, giving me a bath—the pictures all parents take of their children—the two of them look as if they're about to burst. Mom says she didn't even mind when I hit the "terrible twos." I swallowed a bobby pin and had to be rushed to the hospital for stomach X rays. (One aunt, inclined to be dramatic, insisted it was an open safety pin.) Another time, as Mom and I lay sleeping during our shared afternoon nap, I shredded her foam-rubber pillow and mine. "Our children never any trouble," my mother signs, dreamily. "I love taking care of my three babies."

There was one thing my parents did worry about: my learning to speak well. We lived near Grandma Nellie and H.T., and two sets of aunts and uncles, Bill and Margaret, and Ghlee and Guy. I listened to their talk and I loved chattering to them. But more important, while I was still quite small, Mom and Dad bought one of the first television sets, a box that was mostly speaker and base, with a tiny six-inch screen.

Radio, of course, had meant nothing to my parents, but here was a device with small, moving pictures for them and voices for their daughter. A neighbor showed them how to adjust the volume knob, and each time they turned it on, they would feel for the delicate click, then turn to the precise volume for me. When I got a little older and they got a set with a bigger screen, they'd turn the dial, watching me. If I wanted the volume increased, I held my palm open and flipped my hand up. If they turned the knob too quickly, I'd clap my hands over my ears, prying one away to gesture "down."

And it was there, sitting in my small green rocking chair, that I would watch the screen for hours on end, developing what was for northern Indiana a strange, accentless "national-speak," occasionally rushing off to tell my parents about the Cold War or the baseball results. Though our neighbors talked with a hearty midwestern twang, substituting "warsh" and "git" for "wash" and "get," my words were precisely formed, even a little clipped. I spoke rapidly but my voice remained soft. In grade school, it was nearly inaudible.

When Mom and Dad talked to each other, they used American Sign Language. With their hands, they were deft and fast as lightning. Talking to me, they used their voices—throaty whispers—and signed simultaneously in straight English. They tended to finger-spell a separate sign for each letter of a word (a somewhat laborious process), rather than sign (where one symbol can have many meanings), reasoning that spelling was less confusing for a hearing child. As a direct result, my spelling was quite good. But when my parents got into a room with their deaf friends and the signing flashed furiously, it was as if I'd been taught nursery French and then been taken to La Comédie Française.

Speech was always an effort for Mom and Dad, so unless they were talking directly to me, they didn't talk at all, except for those occasions when Mom was teaching me signs. She'd prop me up, point to herself and say "I," hug her two shoulders, "love," then point at me, "you." As it turns out, the first word I used was a sign—"apple"—a sort of inscribed dimple, the joint of the first finger twisting in my cheek. The spoken words followed soon after.

One of the few studies done on hearing children of deaf parents dwelt on one minor but fascinating point: By the age of two, hearing children perceived their parents' deafness well enough to know automatically that they must use gestures with their parents and other deaf people. If the children talked at all, their voices had an unusual quality and they exaggerated their mouth movements. These same children immediately shifted gears, speaking in "normal" voices, with hearing people.

Intuitively I knew not to talk to my mother and father. I gurgled to them to make my lips move the way theirs did, I'm told, but I never spoke distinct words unless I accompanied those words with a sign.

Even today when I sign to Mom and Dad, I make unusual distinctions. If I'm alone with them, I sign and don't use my voice at all. If a hearing stranger is with us and I'm translating, my English is clear. But if I'm with my sisters or an old friend, I use a strange hum-like voice as I sign and as I interpret what Mom and Dad are saying. The voice is almost an unconscious parody of a deaf voice. It seems I only use that voice, though, when I'm comfortable having my parents and certain hearing people in the same room. It's my way of including someone in the family.

The relatives all seemed to breathe a sigh of relief when I was born hearing. (While I was still in the cradle, several of them performed surreptitious hearing tests on me. Sneaking up to my bassinet, they'd carry out the same tests Dr. Douglas had in the delivery room.) Although my grandma Nellie had declared that the deafness in the Walker family was not inherited or inheritable, Uncle Garnel had never had children and thus the genetics of the situation hadn't actually been tested. After all, it was a quirky coincidence, having *two* sons deaf from infancy in one family.

However, Mom and Dad certainly never thought of it as a genetics experiment: They were creating a family. This was the first opportunity they'd had in their lives to hold on to something that was their own and that was permanent. It was an affirmation of their capabilities as people. And the things they'd created—their marriage, their home, their baby—were whole and perfectly formed. They'd suddenly, magically, brought life to their own world. Their home was a haven. When they were there alone, it was exactly the way they'd dreamed it would be.

My parents bought a small white house with a yard and big trees. The house needed repairs and they enjoyed working together on it. They didn't have much furniture, but everything was clean and neat. And they got a dog. Poodie was a darling, honey-colored cocker spaniel. Not only was he good company, but he was helpful too. By looking at Poodie, my parents automatically knew when someone was coming to the door, and at night they depended on Poodie's bark to scare away burglars.

As my parents' child I was part of the deaf world with its culture, even though I could hear and was thus wired to normal society. My parents' friends felt I was "one of them," and years later, when deaf people asked me how I learned sign language, I could see them relax, I could see the relief on their faces, when I told them my parents were deaf.

As soon as I began to sign and talk, I became my parents' guide. They'd been depending upon me for clues even when I was tiny. If I cried though all my biological needs were met, they might realize it was thundering outside. Other times they'd watch my face and eyes, notice that I'd responded to something, and then turn around to see someone walking into the room behind them. Mom and Dad never tried to put pressure on me to do things for them. That just came about naturally. If a phone call had to be made, I was the one to make it. And my being able to do so provided a tremendous convenience to my parents. Before, my father might have to drive hours to pick up a document that I could order over the phone in ten minutes. Yet Mom and Dad never took me for granted. "Please, would you mind to . . . ?" and they'd make the request. They hated having to ask for so much to be done for them. But as practical people, they had no choice but to rely on me.

"You help us to know what happens," Dad often signed of Kay, Jan, and me. "We find out what is interesting. We learn through you."

My grandma Wells remembers me as a three- or four-year-old telling my parents about polio outbreaks and vaccinations.

I have often wondered whether in some ways it might have been easier for them not to have had a hearing child. Certainly it was useful; we were a self-contained family unit. Yet the fact that I was hearing and they were deaf was held up to them every day of their lives. Here were people—their own relatives—coming into their house and acting delighted that the child was not like the parents. "Oh, what a relief you're not deaf," ladies gushed. Well, yes and no for Mom and Dad. Perhaps they would have had children they understood better if we had been more like them. Having hearing children meant they had to come into contact with hearing people even more frequently than they would have otherwise—hearing teachers and scout leaders wouldn't have been such a part of their lives.

"Do you think your parents are jealous you can hear?" a businessman once asked me. It seemed like one of the hundreds of inane questions I'd heard; his frame of reference was so different from ours. "No," was all I could answer. There was just no point in explaining.

There were a few times when Kay or Jan was giving a speech in public, or when I was talking in front of a group, that I saw my mother sign, "I wish I could hear what my daughter is saying." The sign for "wish" is a kind of hungry, longing sign, the hand open and sliding down the chest. But the intention had nothing to do with the differences between us; it had to do with a mother and her daughter.

Still, we encountered strange reversals. When I was in kindergarten, strangers would come up to my father, start to talk, and he'd point to his ear and shake his head "no." He wanted to let them know he couldn't hear, then he held up his hand to indicate "slow," following that with a "come on" gesture, signaling them to talk. All the while he'd be staring at the person's lips. Immediately the stranger would bend over toward me and ask, "Does he lip-read?" as if Dad had suddenly become as inanimate as a cigar store Indian. And that was the turnabout. Usually people ignore children and talk to adults. In our case, I was the five-year-old head of household and they ignored my father. But those were minor irritations caused by strangers. There were other, worse slights.

I must have been about three at the time of one of the Walker family gatherings. I was the only child there. Mom and Dad kept an eye on me all afternoon, the way they always did. I sat in my lacy dress in the corner, cutting out paper dolls.

"Now, Lou Ann, don't cut your pretty little dress with those scissors," one aunt said.

"Be a nice girl and stop playing with those blocks." It was Grandpa Walker.

"Come over here and sit with me," an uncle said, patting his lap. "Be a good girl, now. You know you have to be especially good and watch over your mother and father."

It was a phrase I would hear a thousand times as I was

growing up. Be good. Be good. Be good. And even now I hear that phrase rasping through my head like a handsaw—pushing and pulling, shoving and twisting through a plank.

Mom and Dad watched me carefully and disciplined me when necessary. All the relatives believed I was well behaved, friendly and eager, although content to be off playing by myself. Yet this scenario took place every time we got together with the relatives.

"Everybody would try to interfere," my aunt Gathel once told me about that and many other afternoons of my childhood. "They'd tell you what to do as though your folks weren't even there. Your folks were watching. They could tell what was happening. It was up to them if anything needed to be said to you."

Even as a toddler I was learning that deafness didn't have only to do with broken ears.

My father's first cousin Velma tells the story of when I'd been dropped off to play with her baby, Ann. I was about a year and a half old at the time, Ann a few months younger.

We'd been playing quietly and happily in Ann's crib, both of us scrubbed and neat. Velma, a superbly efficient mother, handed us stuffed animals she'd made and then turned around to fold laundry. Until I started wailing.

She turned back, to see me holding up an arm with teethmarks in it. She was horrified.

"Bite her back, Lou Ann."

I kept on with my crying.

"She bit you, now you have to bite her back. Teach her a lesson."

Velma held Ann's arm up to my mouth. I demurred.

"Come on. Stand up for yourself." Whereupon Velma, the ideal mother, decided to teach both infants a lesson. She leaned over and bit her own baby on the arm.

Ann screamed, even though it was hardly a bite. Apparently I just sat there and stared. To this day Velma is troubled by the fact that I refused to bite back.

It's hard to know where characteristics are ingrained in babies—in the genes, in the womb, in the cradle. To Velma, I think, this incident was early proof that although I could hear, my personality just wasn't tough enough. To me, the story indicated that, for better or worse, I'd absorbed my parents' habit of turning the other cheek. There would be repercussions.

Montpelier is about as American as you can get. People fly the red, white, and blue all year round, and the place hasn't changed much in the last fifty years. Plunked down in the middle of the cornfields, it has one stoplight, where the highway that whizzes through intersects Main Street. You can still park diagonally on Main.

For years the names on the stores were the same as the ones in the two cemeteries. There was Henderson's Hardware, Flanagan's Five and Dime, and Bontrager's Drug Store, where you could still get cherry Cokes and then reach under the table in the booth to feel decades of chewing gum wads stuck to the bottom. There was a movie theater when we lived there, the Palace, with its pink stucco facade. The movies it showed had been out in Indianapolis for a year already before they reached us. TV and the triple-X drive-in took all the business away, though. The Palace admitted defeat and closed soon after we moved away.

People here are real midwestern pioneer stock, mostly Irish, who had come over during the potato famine, and German. These were the ones who were adventurous and hardy enough to make it a third of the way across the country. When they saw the flat, fertile land, the settlers decided this part of the Louisiana Territory was good enough for them. They were

plain people. The only plainer ones were the Amish who settled thirty miles away from Montpelier, in Berne.

Hardly anyone moves to or from Montpelier. Everybody
knows everybody else's business—which can feel like stran-
gulation sometimes, but it has its compensations. You never
have to do any explaining. You never have to lock your doors.
And a kid can have the run of the place. By the age of two, I
was a freewheeling, independent spirit.

It was 1955 when Mom and Dad had saved enough to buy
the old wooden house on Jefferson Street in Montpelier. Mom
worked on fixing it up during the day, and Dad helped her in
the evenings. Weekends—I was about two—I'd stay with
either Grandma Walker or Grandma Wells while Mom and
Dad did heavy repairs.

One weekend Dad was laying linoleum tiles in a closet when
some of the boiling fixative ran down his leg, catching fire. He
thought my mother was out on an errand and so he ran to his
brother Bill's house down the street for help. No one was
home. He ran across the street to a neighbor, who figured out
what was wrong and called the fire department. In the mean-
time, my mother, who had actually been in another part of the
house, smelled smoke and rushed out. When Dad got back to
the house, a policeman saw him limping, his leg charred. He
rushed Dad to the doctor. Soon after, Bill came with an ambulance to rush him to the hospital. Dad spent months in the in-
tensive care ward with third-degree burns over his legs. Mom
and Dad's house, along with most of their savings, went up in
flames.

Dad underwent skin grafts—sheet grafts made with a tool
like a cheese slicer—without anesthesia, Aunt Gathel told me
years later. The doctors worried about the effects of anesthet-
ics, fearing his lungs might have been burned as well.

Grandma Nellie sat in the waiting room with Doris Jean
during the skin-grafting surgery. She heard a piercing scream

and knew it was Puff. She said she didn't see how he could keep from going crazy; the pain must have been terrible. But she was glad Doris couldn't hear it and Nellie tried not to show how she was aching inside for her youngest son.

Dad has no memory of calling out during the operation.

H.T. had started a funeral business and ambulance service back in 1912. It had gone through tough times over the years. Montpelier, though a tiny town, had several competing funeral homes and H.T. was not a particularly competent businessman. He overstocked expensive caskets, and he seemed to spend an inordinate amount of time on outside projects. During World War II, Uncle Garl would finally help get the business into the black, then he would leave to run another establishment and Uncle Bill would come back from the war in the South Pacific and stay on.

The funeral business sounds more gruesome than it is. It's actually a lot of waiting around—except when there's a funeral, and then everyone is scurrying, putting out folding chairs, carrying baskets of flowers, picking up dresses or suits, chauffeuring family members, delivering death notices and boxes of sympathy cards. People don't die on schedule. Sometimes there are several funerals in a week. Some months there aren't any.

Whenever Puff was home, he helped by rushing to accidents with the ambulance, picking up bodies from other states, and driving the flower car in funeral processions, doing much of the heavy work required. He was also called on to transport sick people between hospitals and drive chairs out to reunions whenever a family wanted to borrow them. My aunts and uncles were impressed with his industry. Over the years he put in thousands of hours.

In the middle of one night, early in Mom and Dad's marriage, the two of them were asleep, with their window shade slightly open. There had been a bad car wreck, and Bill came around to their window, took out a cigarette lighter, and flicked it on and off until Dad finally woke. Dad spent most of the night ferrying bloody bodies from the scene to the hospital, then headed directly to work early the next morning.

Just before the fire, Dad had been offered a newspaper job in Fort Wayne, a town that was then about an hour-and-a-half drive from Montpelier. It was a better-paying job in a larger shop. Dad went in and told his father of his plans, writing on a piece of paper when he was going and where. H.T. put the paper down, closed his eyes, and gave Dad an elaborate shake of the head "no." Dad was crestfallen. He looked into his father's eyes. "Why?" he said in his quiet voice. He held out his notepad and pen to his father for an answer.

H.T. wouldn't take the paper. "Too far," he said aloud. "Stay here at home." Then he pointed at Dad and pointed straight down at the ground. "Right here."

When H.T. didn't get his way, he could make life very unpleasant. Reluctantly, Dad stayed. Both he and Mom knew how hard it would be to move to a new town with a new baby and without any parental support. Looking for a place to live at a time when landlords wouldn't rent to deaf people and setting up bank accounts when banks often required deaf people to get co-signees just for checking accounts wasn't going to be easy.

Then there was the fire. Most of their money had gone to buying and fixing up the house, and with the medical bills and Dad not working all those months, they needed a bigger paycheck. Another offer came, this time from a deaf high school classmate living in Dallas, Texas. The friend had written saying there was an opening at his newspaper, that there was a nice deaf community in Dallas, and the pay was higher. To my father, it seemed like a dream opportunity. In Montpelier they didn't have any deaf friends. Garnel and his wife, Imogene, fed up with their farm, had moved to Anderson, Indiana. Mom and Dad were feeling isolated.

Again Dad went in to announce his intentions to H.T., who told him he absolutely could not leave. H.T. informed Dad that he had obligations to the family business, that the family had helped him out when he was sick and he couldn't go off with a bunch of strangers to Texas. And then he invoked the big guns: "You'll break your mother's heart."

Dad complied. For a while.

Later that year, we all went off to the Montpelier Parade.

There's a black-and-white photo of Dad standing next to his decorated bicycle, looking straight into the camera, beaming with that gap-toothed smile of his. He'd just won first place for best decorations and there are yards and yards of colored crepe paper woven through the spokes in his bike. A banner that says "Welcome Montpelier Day Parade" is stretched between poles attached to the front and back of the bike. He's wearing a felt hat, cut from an old fedora, pulled straight down on his head.

When Dad was a kid, the parade had marked his favorite day of the year. Beauty queens—actually bashful local girls in prom dresses—rode in the backs of convertibles, waving, throwing pencils and small toys and trinkets to the crowds. The convertibles were decorated with pink and white tissue flowers, many of which Dad had made himself. Uncle Bill got out his collection of antique cars, and when Dad was a few years older, he'd get to drive one himself. Every year they had a big raffle. The year I was three, the grand prize was a bike. I won. It was a full-size, two-speed, fat-tired, fiery red Columbia, my pride for my entire childhood. I didn't even know I'd entered the contest. Mom and Dad hadn't said anything to me about it.

As it turns out, I hadn't entered. H.T. had rigged the contest.

First, he stuffed the box with raffle tickets that had my name on them. He was the master of ceremonies that day, as well as the person who drew the winners' names out of the raffle box. Apparently, he hid behind the curtain of the gymnasium stage, throwing away every name he picked out of the box until he finally came to mine, and proclaimed me grand prize winner. H.T. had wanted to please his granddaughter and his son. When Dad heard the story several months later, he made no response. His face went very white; his mouth was set. It was a turning point. He resolved he would not subject his family to domineering any more.

My sister Kay was born a year and a half after that. It had been astonishing to watch my slim mother inflate like a beach ball all those months. One day, several weeks after my fourth birthday, she brought home a chubby new baby.

There was no question of sibling rivalry. Kay was mine. I helped diaper her. If she cried during the day, I rushed to get Mom. I watched her and held her as often as I could. When she was awake, I'd go in to her room and talk to her, leaning my forehead on the bars of her white crib. She was a colicky baby who often cried, and that was the only thing that displeased me—having the accustomed quiet of the house disturbed. I'd walk into her room and whisper to her, trying to calm her, trying to jolly her up, my small hands clumsily patting the top of her head, cooing to her, the way I imagined mothers did for their babies. I was tall and skinny for four and I loved being allowed to push Kay's stroller. Kay was an extraordinarily shy baby who hid behind my mother's legs, clinging to her skirts.

"Now you look after her," a cousin would tell me. "You talk to her a lot because your mother and daddy can't." Soberly, I heeded these instructions. As we got older, Kay and I almost always played together. If we were outside in the sandbox—a tractor wheel turned on its side and painted white—Mom would sometimes go inside and do her housework. She'd glance out at us as we filled pails with sand. If the sand was wet from the rain, we'd build castles. Rather, I built the castles and Kay smacked them down.

Our temperaments were as different as our coloring. She had pale-blond, almost whitish hair, and although she was fair, she would later tan. Her chin was nicely rounded like Mom's; the towheadedness came from Dad. I don't know what she was thinking about it all, but I was flourishing, playing teacher, encouraging her to speak, instructing her in the arts of sand-castle building, singing, and everything else I could think of. It couldn't have been that easy for her—power is a heady thing for a four-year-old—but to everyone who saw us, we seemed as close and happy as two sisters could be.

I'd just turned six when Dad drove down to Indianapolis to the *Star* and *News*, two sister newspapers, and handed his union card to the foreman of the linotype room. He was hired on the spot. He drove back to Montpelier, told everyone he'd taken a new job, that there was no turning back. He gave his boss notice and in February of that year he moved to a rented room in downtown Indianapolis. After work he went house hunting, driving all over town to find a neighborhood he liked, one that was clean and good enough for his family. In the spring he found a half-built frame house in the suburbs to the east of Indianapolis.

H.T. complained to his other children that he couldn't understand why Puff would up and move away from his family like that. Gathel suggested to him that Puff was thinking of his future. "Why don't you try offering him a part of the business?" H.T. said that he had and that Puff had turned down the offer. That was not true. No one had ever asked Dad to be a partner in the funeral home. Nor had they offered him the assistant's job they gave to outsiders. No one had even offered to compensate him for all those back-breaking hours he'd spent carrying caskets.

I was excited about going to a new place, but the move would prove to be momentous and difficult for all of us. We were leaving safety, the place we knew and where we were understood, for a big city full of strangers. At first we felt all alone. We had no relatives living there. Slowly Mom and Dad began reestablishing contact with the friends who'd gone to high school with them. Dad's job was much better and my school was excellent.

Before leaving Montpelier I'd been serious and self-conscious enough, but this new place, as placid and ordinary as it seemed, was overwhelming. If a neighbor came by, I would make the introductions and explain where we'd come from. But mostly people seemed to circle our house, eyeing us. I could see the window shades lifted and the dark shapes of heads in the windows whenever we walked into our front yard. In Montpelier, people hadn't been that way at all. Gradually, through the neighborhood kids' responses to my parents—the ones who weren't shy about staring open-mouthed at Mom and Dad—I figured it out. These people had never met a deaf person before. We were a curiosity. For the first time, it hit me that my mother and father were deaf.

It was a gray afternoon in early spring. The yards were too muddy to walk in. I saw some kids playing in the driveway across the street from ours and suddenly I decided I should go meet them. I figured the easiest way would be to bring over some kind of present. I went in and asked Mom for some cookies to serve.

"I want them on this plate," I told her, climbing up on the counter to retrieve one from the cabinet. The dish had a delicate blue and yellow floral pattern on it. Mom had splurged on a set of dishes to go with our new house. They were called Melmac, a tempered plastic, guaranteed not to break. The guarantee was what intrigued me.

I took out the Oreos. The kids just looked at me. Finally, they took the cookies and as they stood holding the dark wafers in each hand and licking out the cream centers, I announced that the fragile-looking plate was unbreakable. Naturally, one of the kids was skeptical. I let him smack the plate on the asphalt street. The first several bangs didn't do it.

"So what," one boy said.

"It won't break no matter what," I told him. "You should hit it harder," and I took it over to the large rock at the edge of the drive, held the plate up over my head, and brought it down with a smack that left a chip as jagged as a broken tooth.

None of the kids said anything. I felt more like an outsider than ever. I walked back into the house, scared Mom would be mad. She was so proud of her brand-new dishes.

"I broke your new plate," I confessed, "broke" signed by putting my two fists together and twisting them apart.

"I know. I saw from the window," she said. "It's all right."

Instinctively, despairingly, she knew what the real problem was that day of the Oreos, and it was one she couldn't solve. She got down on her knees and put her arms around me, then she held me away. "You take things too hard," she signed.

I never had the feeling I was being watched, but from one window or another, over the years, Mom would see me fall off lawn chairs and bicycles, drive the car into ditches, and neck in the driveway. She watched and knew I had to get into and out of my own scrapes. She knew I was too independent to allow for any interference. She also knew there wasn't much she could do.

On a sunny morning that same summer, a neighbor, the mother of someone I was going to play with, came out of her house. She was a pleasant-looking woman, wearing a dress and an apron.

"Are you the one with the mother and father who are deaf?"

"Yes."

"Oh, you poor little thing. It must be terrible. Them both being mute that way."

She was nice but she was feeling too sorry for me.

"No, it's okay." I wanted her daughter to hurry up so we could go play.

"Can't they talk?"

"Yes, they can, but their voices are sort of hard to understand. I interpret for them. We're fine. Really."

Later on that day, one of the kids I was playing with said, "It must be great, your mom and dad can't hear. You must get away with murder. You can yell all you want and they never scream at you or nothin'."

"Well, it's not so great. It's hard for them to be deaf. And I don't scream much and they can yell at me if they want."

I writhed under the scrutiny of outsiders. I didn't think they ever got it right, how it was in my family. They thought it was either better or worse than it was. I never wanted to go on explaining too long. There was too much of it to do.

I hardly ever brought kids home with me. Mom asked me not to, saying the house wasn't straightened up enough—in fact, it was *always* neat and clean. The real reason was that in this new place among strangers, her house was a refuge. It was easier having her daughter run small errands, talk to door-to-door salesmen, and act as her buffer to the outside world.

Listening

7

Spies

I'm not quite sure when I decided they were spies. At one point or another, most children think they're adopted. I never saw it that way. I was convinced Mom and Dad had been sent to check up on me. At about the age of eight I wasn't politically savvy enough to think about who might have dispatched them to me—or why. But there was one thing I was sure of: Mom and Dad weren't really deaf. They were pretending not to hear so that they would know everything I was saying and denounce me for it.

The tests I devised were simple and fairly ingenious. One of my ploys was to go into another room and scream, "Mom! Help!" Nothing. Then I'd drop a book on the floor and hold my breath. Still nothing. I'd throw myself with a thud onto the carpet. Nobody came to check.

Other times I'd sit on the floor in the living room and watch them while they read the paper. I wanted to see if they would make a mistake. (Maybe they weren't as well trained as I'd been led to believe.) I'd sit for long periods watching the back of my mother's newspaper. When she brought the two sides together to turn the page, she'd catch me staring at her.

"What's the matter?" she signed, her forehead wrinkled.

"Nothing."

"Why are you looking at me?"

(The one thing that completely unhinged her was someone staring at her—at home, in a restaurant, anywhere.)

"Oh, sorry, Mom. I was just thinking," I signed, index finger circling my brow.

She'd go back to her newspaper and peer around the corner a couple of times, not able to figure out what I was up to.

While they were sitting in their easy chairs in the living room, I'd put my hand over my mouth, lower my voice, and announce, "Doris, your shoe's untied." Not a flinch.

"Gale, Soviet troops have just crossed the border. Red alert!"

When those tricks didn't work, I decided to tail them. I'd walk up the stairs behind Dad when he was going to his bedroom. Then, as he headed to the right, I'd make a wide circle behind him, maybe even diving under the bed so I could watch his ankles. He usually went to his bureau to get a handkerchief, and I knew it was useless to search there. I'd watched my mother put clothes into the dresser drawers hundreds of times.

Mom was harder to follow. Even if I ducked into a room, she'd catch my reflection in a hall mirror. Her sixth sense was keenly developed and she knew when someone was standing behind her. Usually she was amused by my game. But if I kept it up too long, she'd become exasperated.

"What are you doing? Go play. Go outside." And she'd gesture toward the front door.

One night before I went to bed, I fixed the phone cord just so. The next morning it was in exactly the place I'd left it. And the next. And the next.

Years later as an adult, during an afternoon of reminiscing, I embarrassedly confessed my delusions to Kay and Jan, only to find out that they, too, had both been convinced Mom and Dad were spies. Each of us, in our turn, had concocted nearly identical tests for our parents.

Although signing is not a pantomime of daily action, there are signs that are parodies of actual objects and that retain a cleverness even for deaf people. Indeed, these signs are to sight what onomatopoeia is to sound.

"Watermelon" is an example. You thump the back of one hand as if you're testing the ripeness of a melon. For "banana," you peel your index finger. "Monkey," you hunch over while scratching your sides. For "ice cream," you pretend to lick a cone. "Spaghetti," your little fingers wind around each other, then you draw them apart, still circling, an imitation of the long, thin, slippery strands. In "elephant," your flat hand starts at your nose and then snakes out in the shape of the trunk. For "onion," you put the knuckle of your index finger at the corner of your eye as if you were wiping away a tear. With "turtle," one hand—the shell—covers the other, and the thumb peeks out and bobs up and down as if it were the turtle's head.

One day Mom went up to Jan and signed without moving her lips: "Do you want some . . ." and then she did a little rumba step and shook an imaginary round object at each side of her head.

"What's that?" Jan asked.

Mom did the rumba again.

"Coconut."

"Why is that coconut?"

"Feeling the milk inside," Mom said, but this time she signed "milk," something we usually spelled out.

"What's *that*?" Jan asked.

Mom repeated the sign for "milk," a sign made as if she were actually pulling the teats of a cow, milking.

"Gross!" Jan said.

It was Mom's turn. "G-r-o-s-s. What's that?"

Jan made a face. "Yuck."

Mom returned the face. "That's nature."

Kay and I had been sitting on the sofa watching this amusing exchange.

"You know," Kay whispered to me, "Jan's a great kid."

"Yeah, I know what you mean," I answered. "She's really normal."

Both of us were seriously engaged in a discussion of Jan's attributes, of how well she got along with everyone. I'd once read that each child is born into a different family. Jan certainly was—one where there was more talking going on around her.

Jan's was an ease both Kay and I longed for—especially Kay, who was so deeply cautious and whose shyness made her especially uncomfortable around outsiders. She was a middle child, and my role, as not only eldest but also interpreter and dealer with the rest of the world, made us very close. I was the one Kay asked questions of when we were growing up because I was the one who had the most contacts with the world. Kay and I played together most of the time, but in addition to being the explainer, I was also quite protective, and that must have been hard on Kay. I enjoyed playing teacher, helping her with her schoolwork, but then when Kay wanted to feel protective and grown up, she turned to Jan. And Jan was too independent for that. There's still a dent in the closet door of the room they shared. "Stop being so bossy!" Jan had yelled as she hurled a shoe across the room at Kay, who ducked just in time.

Jan was not only the youngest but also quite petite, and Kay in particular was fond of a photo of Jan sitting on our front porch, wearing a red dress and plaid jumper. Jan has one finger at her chin, just under the dimple, smiling at the camera. Her straight-across, blunt-cut bangs are a little ruffled from sleeping on them funny.

As young as she was, Kay could recognize that Jan was right in her estimate that her two older sisters worried too much, that we were too serious, too wrapped up in grades. Periodically, we would have that same discussion over the years. "You know," Kay said, "Jan's got common sense. She's cut out for the long run."

My ritual every night as a girl was to stare up at the patterns the curtains made on the walls, the moonlight reflecting through the tiny holes in the eyelet lace. The yellow-gray figures danced on my walls as clouds passed over the face of the moon or as the breeze rustled the trees. As I lay in bed waiting for sleep to come, I'd listen to the noises outside. That was what pure hearing was: straining to detect a rustle, a distant motor, a siren. Back then I used to wonder if Mom and Dad had ever tried reaching out and grabbing noises that way. Or whether it was for them the same sensation I felt after the katydids stopped their scratching: absolute silence. And when the katydids remained quiet for too long, I'd move my head against the pillow or brush the sleeve of my pajamas along the sheet—just to hear something.

Once asleep I was as good as dead, except for those times a crash awakened me. I'd scramble to the sill and lift the shade to peek out. It was always the same source, the widow next door, Mrs. Haymaker, slamming her windows against the evening chill. I'd lie back in bed, tucking the sheets around me, folding my arms across my chest, determined to sleep. Only I was really listening and watching.

One night, though, the summer I was eight, I'd gone to sleep, been awakened by Mrs. Haymaker's window slamming shut, and was just drifting off again when I heard scuffing noises next to my room. It sounded as if someone were knocking on the house, on the wall right next to my bed. I sat bolt upright and listened. I was sure I heard a man's voice. He said a couple of sentences, but I couldn't make out the words. I waited a long time, poised in the dark. The noises didn't go away, and I was too afraid to open the shade and look out. Every muscle in my neck was strained as I tried to hear what he was saying and what he was doing by my house. I lowered myself out of bed to the floor, careful not to let whoever it was see my shadow, then I crawled to the door and rushed upstairs to wake Dad.

Lightly I put my hand on his shoulder; I didn't want to startle him. He didn't move. I tapped him. He still didn't budge. I tugged on his arm until he finally woke up. It was too dark for him to see what I was saying. He squinted at my hands, then touched my hip to move me over to the moonlight coming through his window.

"Man outside my room," I signed. I pointed in the direction of my bedroom. "I heard a man."

Dad threw on his robe, found his glasses, then grabbed the flashlight he kept in a drawer. I was following close behind as he lumbered down the hall, so sleepy he ran his hand along the wall to keep his balance. He went through the living room and peered out the front door. He flipped on the yellow porch light and walked out to the front yard as I stood behind the screen door. He looked eerie in the glow of that light, heading around the side of the house while I crept out to the edge of the front porch. As he walked along the corridor of grass between our house and Mrs. Haymaker's, I ran to the corner and peered around. He'd been swallowed up by the dark. It was the middle of July. I was barefoot and shivering.

Eternity passes when you're waiting in the dark, listening for violence. Dad must have gone through the entire backyard step by step.

When he came back, relief flooded through me. I'd felt so vulnerable, standing alone in the front yard, no cars passing, no lights on in the neighbors' windows.

"Nothing. Sorry." Only my father would apologize when someone roused him in the middle of the night for nothing.

He patted my shoulder and we both went back to bed.

I heard the scuffing noise several more times that summer and the one after and the one after that. I'd wait for a long time, listening to the noise, trying to make sure that something was really there. A few times I didn't make Dad get up. The other times, the scenario was always the same: Dad went out, me trailing behind, looking, listening, shaking. He'd search the whole yard, then walk to the end of the driveway, flashing the small beam up and down the street.

He never once refused to go out, nor did he ever tell me I was making it up, but after I'd awakened him several times that third summer, I decided it must all be in my head. I vowed not to bother him again.

It was the end of that summer and I was in bed having an open-eyed fantasy, watching a crack slither around on my ceiling, feeling all the different ways I could swallow, when I heard the scuffing again. I didn't go get Dad. I heard the man's voice quite distinctly that night, but I tried to convince myself I was imagining it. I couldn't face the embarrassment of having Dad troop out and find nothing all over again, and I hated sending him out defenseless in the dark. I figured all I had to do was talk myself out of it. I waited awhile, tense, alert, even though I didn't know what I would do if my imagination actually cut through my screen window and came into my bedroom. I waited awhile longer and the noise stopped.

And then I heard a shotgun blast.

I tore out of bed and ran upstairs, this time waking both Mom and Dad. The three of us rushed to the backyard and there was Mrs. Simon, our neighbor, holding a real, live smoking gun, shaking it in the direction of a mustard-yellow house down the street. In the commotion, the whole neighborhood rushed outside. At first I couldn't figure out what

was going on. Finally, Mrs. Simon grabbed the cigarette from
her lower lip, pushed the strands of coarse hair away from her
bony cheeks. With the smoldering cigarette held between her
fingers, she pointed at me.

"Next time I'm goin' to shoot him. No warning. I'm goin' ta
shoot him."

I signed it all to Mom and Dad. We were perplexed.

"What is it?" Dad asked me.

"I don't know. I can't understand," I signed back to him.

"*Who* are you going to shoot, Mrs. Simon?"

"He was pretending like he was walking that big dog of his,
but I know what he was really doing. Snoopin' around, nosin'
around other people's property, looking in windows."

"*Who?*" I practically begged her.

"Hessel, the one who lives over there in that house with the
brown shutters, the one with the dog he calls Hessie."

Mr. Hessel, who stood well over six feet tall and must have
weighed at least 250 pounds, kept a beautifully groomed Saint
Bernard. It was true that whenever he walked the dog, this
enormous man acted as if the dog were walking him. Hessie
was the perfect unwitting accomplice: He was so well behaved,
he never barked when the two of them were walking among
the houses.

I didn't think Mrs. Simon was much saner than Mr. Hessel.
It was unsettling knowing there were guns and Peeping Toms
about, but still I was relieved. I no longer had to mistrust my
hearing. If anything, it was too acute.

My father's determination is a funny thing. He never an-
nounces resolutions or gets frustrated, the way most people
do. There's no braggadocio to him. Whenever I went to him
with a broken toy, he would take it in his strong, handsome
hands, the hands he washed so thoroughly each day to get off

all the printer's ink, then study the toy carefully before doing anything to it. He'd nod, carry the toy to his workbench, and no matter how long it took, he kept at it until he finished the job.

When I told him my desk chair was wobbly, he came into my room, looked it over, tipped it up, then stood back and rubbed his chin. As the solution came to him, he'd point in the air, and sign, "Yes, I can do," and he'd get his tools. Sometimes I thought I had a better idea about something and I'd tap him impatiently on the shoulder. He'd turn to look at what I had to say and nod. If he could use the idea, he would. If his own method made more sense, he'd tell me to watch a little longer. If I interrupted him too often (or tapped his shoulder so hard it hurt), he'd say "Wait!" aloud. He was a master craftsman because he had spent so much time watching others, and because he was a patient worker. Occasionally I helped him as he repaired the car, listening for him until the motor hummed just right. But for the most part, he knew exactly what to do with an engine by using his eyes, his reason, and by being determined to complete the job. Yet no matter how determined he or my mother was, and despite the fact that they are terribly bright, there was one thing that eluded them: English. I was about eight the first time Mr. Hessel came to my window and it was the year I began feeling very adult. Mom and Dad started asking me to correct their letters.

Writing was hard for them, even harder than reading. In writing, you can't gloss over things the way you can in reading. A conservative survey once showed that the average deaf high school graduate has a third-grade reading ability. Writing is more difficult to judge. I only know that each time Mom and Dad had to fill out forms, they read slowly and carefully so as not to make any embarrassing mistakes. If I was there, they'd tap me on the shoulder and ask what a word meant. My starting school, of course, forced them to face dozens of forms and I could tell Mom despaired whenever she had to confront the papers with medical histories, family background, or "In Case of Emergency" written at the top. If a line said "previous domicile," Mom would turn to me and say, "What's that?"

"Where you lived before," I'd sign.

"Should I put Montpelier?" she'd ask, her faced turned up almost girlishly as I stood beside her at the table. She was so afraid of making a mistake.

"Yes, Mom."

She'd finish that one and turn to the blue paper. "What do you think I should write for 'Emergencies'?" We didn't have a phone then, and even if we had, she couldn't have answered it. Mom sent me over to Mrs. Miller's house across the street to ask if it was all right to put her phone as our school emergency number. From then on, I wrote the Millers' number and my uncle Bill's number in Montpelier, just in case, and it was Uncle Bill's number that I carried with me in my wallet, even though he lived a good two hours' drive away.

Letter-writing held the greatest challenge for Mom. I watched her struggle with that first letter she was going to ask me to correct. She approached it as if it were a formal composition, taking out a large, clean piece of paper and a blue-ink pen, and sitting down at the kitchen table. She smoothed the paper, twisted the pen in her fingers, and licked her lips, all in preparation. Finally, she wrote the date and the greeting, and as she continued with the letter, she'd occasionally stop, pen poised over the line, reading and rereading what she'd written. When she finished, she added her name to the rough draft. Then she handed it to me. "Will you fix it? Make it sound better," she signed. For "sound" she pointed to her ear.

There were a few cross-outs in the simple sentences she'd written. Over the years, I could visualize where she'd learned the constructions she used. Some were from her school-day grammar books. The books had always shown the infinitive form—"to play," "to work"—and when Mom used verbs in her letters, she often included the "to" when it wasn't necessary; for example, "We enjoyed to call her."

Other constructions came from sign language syntax. She would have an adjective following a noun rather than preceding it—"dress red" rather than "red dress." Much of my work was to make the letter conform to the receiver's expectations. I would add more formality to a business letter or unstiffen a personal note.

After I made my copyediting marks and handed it back to her, she'd thank me, her open hand pressed to her mouth, then courteously she would arch the hand toward me.

"No problem," I told her. I'd watch her take the sheet and read it over, pinching her lower lip between her fingers in deep concentration. She didn't just want to copy it over; she wanted to learn from the exercise.

Dad and I went through the same routine. And later Kay and Jan began correcting letters as well. In fact, even if Mom and Dad were sending a letter to *me* they'd have one of my sisters correct it, and it became a game trying to figure out whether Kay or Jan had worked on a particular letter.

Mom and Dad's sentences sounded as if a foreigner had written them, as if English weren't their native tongue at all—and of course, it wasn't really. Still, I was to find out later that their writing was far superior to most done by deaf adults—even deaf college graduates.

Back then, a proud and self-conscious third grader correcting her parents' letters, I was filled with a mixture of pleasure and embarrassment—pleasure because I was useful to my parents, embarrassment because they couldn't do what I thought all other parents did with ease.

One morning that year, after I'd been sick, Mom tried to write an excuse note for me. She'd been rushing around the house, making the breakfast and beds, helping everyone get ready for the day, tending to Jan, who was a baby then. She sat down at the table but kept crossing out the words on her paper.

"Here, you think it up," she said in sign, and handed me the pen. I wrote it and she then copied what I had written. It was amusing in a way to think that I was writing my own excuse notes while other kids were forging theirs.

Every day after I left for school, Mom would lock the doors around the house—a wise procedure since she wouldn't hear if an intruder entered—and every day just before I came home from grade school at 3 P.M., she'd unlock the front door. Actually, I wasn't even aware that this was her habit until the one and only time she forgot. That day I'd gotten off the bus at the corner and run home because I was desperate to go to the bathroom. The door was locked when I arrived. Standing on tiptoe on the front porch and leaning way over toward the window, I could see Mom's back as she mixed a cake in the kitchen. She was intent on the cookbook and the cake. I banged on the door, on the window, I waved my arms wildly, hoping to make a shadow. She didn't look up. I ran around to the backyard. The dog was barking like crazy as I hammered on the doors, looking for a window that might be open, but it was winter and everything was shut tight. I kept running around the house, trying to get in. After what seemed like an eternity, Mom glanced at the clock, started, and ran to the door to open it. I rushed past her. Immediately she knew what the matter was. She came running behind to help. It was too late. I was wearing French-blue tights, my very first pair. Standing outside in the cold, pounding on the door, I was angry at Mom for having forgotten me, furious that I could see her but couldn't get her attention. I was ready to complain bitterly. Then I looked up at her woebegone face. I saw that she felt worse than I did.

things hearing take for granted

* * *

If Kay or Jan or I had been wilier, we could have presided over anarchic terror. But it wasn't that way at all. There was an orderliness, a gentleness to our lives. Mom and Dad maintained the internal rhythms they had developed at school. We ate meals at a fixed time—dinner was at five-thirty, except when Dad was working the night shift, and then, even though it was four o'clock, we all ate together. Friday nights we ate out. Saturday morning was for laundry. Once a month, on a Sunday, we drove to visit my grandparents in Greencastle.

Although my sisters and I still had our suspicions about our parents and their abilities as secret agents (we hadn't yet caught them red-handed), Mom and Dad's parental control was simple and direct. Mom says we required very little discipline, and there were never any double messages from our parents. If Mom discovered one of us marking in a book or coming home late from school, she'd say, "I don't like that. Nice girls don't act that way." If Dad was really provoked, he'd give a swift flick with his thumb and middle finger to the side of our heads. "Not nice," he'd sign. And we were chastened. Above all, we wanted to be nice. That's what the aunts and uncles and grandparents and school teachers told us to be. Kay and Jan and I tried to call as little attention to ourselves as possible. We wriggled when we were under the spotlight. Public attention seemed a terrible thing. It was like being stared at in a restaurant.

There is a subtle tyranny in being nice, in worrying about what other people think, in being concerned about how everything you do is perceived, in feeling that you're constantly scrutinized. That tyranny is what makes schoolboys have bloody fights in their Sunday best. It's what makes little girls into tattletales.

Mom and Dad were instinctive parents, parents who imparted a strong sense of right and wrong. They were easy to live with and we wanted to be nice with them. But the outside world was confusing. We wanted to be nice and not attract attention, but we were also desperate to make Mom and Dad proud of us and we needed to be proud of ourselves. Kay, Jan, and I worked hard on Science Fair and art class projects. We won essay contests and spelling bees and music awards.

It was fourth grade with its flickerings of puberty that proved to be troublesome—because of and despite the fact that I was such a good girl. People expect us to act in certain ways and it's the surprise that disturbs them. When the class bully behaves for an afternoon, the teacher is suspicious. When a docile child acts up, the teacher overresponds. I used to sit at my school desk, back straight, hands folded, paying strict attention to whatever it was the teacher wanted. I tried to keep all the teacher's admonishments in mind: I didn't talk; I kept my hands to myself; I did what I was told. I used to sit next to the class clown. One day I failed a pop quiz. I stung with humiliation, which was bad enough, but then the boy teased me about it. At first I ignored him. But he kept it up until finally I couldn't stand it anymore. I pinched him. He howled and told the teacher what I'd done.

"What's the matter with you?" she barked. "You can behave better than that! Now stop it and be a nice girl."

Mrs. Wamsley was the most hated and feared teacher in all Moorhead Elementary School. Her left leg was nearly a foot shorter than the right and she had a large, squarish body which she dragged slowly through the halls. Keeping up with thirty wriggling ten-year-olds every day must have been painful for her. She always seemed physically relieved whenever we were all quiet in our seats and she could get a few seconds' rest. She'd stand at one side of her desk, hand holding the edge, as her left leg and black platform shoe dangled in the air.

She made lots of rules. If you ate a school lunch, she wouldn't allow you to turn in your tray until you'd taken three large bites, even of the soggiest green beans. If she was mad at one kid, she might keep the entire class in from recess, heads on our desks, as punishment—though school rules expressly forbade that. Oddly enough, I always felt sorry for her, but I was mortified when one of the other kids accused me of being Mrs. Wamsley's "pet."

Mrs. Wamsley's most oppressive rule banned talking except when we were reciting. To circumvent the rule, we tried note passing—entire notebook pages, covered with writing, were specially folded into two-inch squares, with a triangle folded over the top.

In the row next to mine sat an enterprising girl, Vicki Terrell. When she discovered I knew signing, she thought she had a perfect way to get around the no-talking rule.

"TEACH ME THE ALPHABET IN DEAF-MUTE LANGUAGE," Vicki wrote me.

I read the note, glanced up to see if Mrs. Wamsley was watching, then shook my head "no" at Vicki.

"Come on," she hissed. "It will be fun."

"I don't want to."

"Don't be a baby. Teach me at recess. What could it hurt?"

I couldn't articulate it then, but I would realize later that what she hurt was my sense of privacy. Never once have I failed to feel a pang when asked to show some signs. It seems like too public a display. Somehow it trivialized us, me and my family, making the way we talked into a party game.

Despite my lame arguments, Vicki shamed me into teaching her. (That was one of the things about deafness. It made us compliant, unable to resist authority.) Vicki had broad, flat fingers that gave her trouble, but by that afternoon every girl in the classroom knew the alphabet forward and backward. Secrets got passed for an entire week. I'd occasionally correct a girl if she couldn't remember how to make an *x*, but I rarely signed myself. Mostly I'd nod if someone signed to me. By

Monday some of the boys had caught on. Security became lax.

"Pam! What are you doing?"

"Stop that this instant! What was that?"

There were some embarrassed coughs in the room as Pam straightened up in her seat and tucked her delicate white hands under the metal desk.

I could only be grateful that I hadn't been singled out, but I felt bad for Pam. Somehow it was my fault she'd got in trouble.

That year Moorhead School was making a big parent-relations push.

"The teacher says you can come visit my class," I told my mother one night. I handed her a mimeographed form. In a way I didn't want to ask Mom. It seemed too public, too uncomfortable, but the teacher had made such a big deal about the whole thing that I thought I had to.

"You mean spend the day at school?" she asked.

"Yes. Don't you want to?"

"Well, I don't know what I'd do. I can't hear. . . ."

"That's okay. Don't you just want to come and watch?"

Against her better judgment, Mom was talked into it. She found a baby-sitter for Jan and came for a morning. She looked so pretty, dressed up in a light-green suit and high heels. And as soon as she arrived, I knew it had been a mistake. She was the only parent who came. The teacher seemed to act as if it were an imposition having Mom there. She did little to welcome her except shove a chair at her and point toward a spot.

Mom sat and watched kids writing on work sheets and raising their hands to ask questions. I could hardly look at her. I didn't interpret for her at all. Every once in a while a kid would turn around and stare at her, then at me.

Lunch was worse. Mom stood in line with me—she wasn't even accorded the privilege of most guests, going first and sitting at the head of the table with the teacher. By the time we got to the twelve-foot-long tables and benches, the only two spaces left were smack in the center. It was fine for little girls in loose dresses to hoist a leg over the bench. It was completely impossible for Mom, wearing a straight skirt and heels. I watched her place her tray down and totter, trying to climb over.

"Should I ask someone on the end to move?" I signed in the middle of her maneuver. All eyes were on us.

"No, it's fine," she signed, taking a steadying hand away from the table. "Don't worry."

She left after lunch. Mrs. Wamsley walked with Mom and me to the door and into the hall.

"I wanted to tell you what a fine young lady Lou Ann is," she said to Mom. I was completely taken aback by her compliment, and writhing because I had to sign that kind of thing. "I'm pleased to have her in my class."

"Thank you," Mom managed aloud, then she gestured toward me, nodded, and flashed her most beaming smile, conveying to Mrs. Wamsley that she was proud of me.

By example Mom was trying to show me that it didn't matter that she was stuck in the back of the room. That the family was what was important. But to me, it mattered.

The big excitement of the year was getting a telephone. It was an ivory-colored wall model, hung up in the hall next to the kitchen. I was thrilled. I felt about it the way the other girls in my class felt when they got their first pair of stockings. To me, having a telephone in our house meant real sophistication. Installation and monthly bills were expenses Mom and Dad wanted to avoid, but it had become too inconvenient to drive all the way to the dentist's or the doctor's office just to make an appointment. I was delighted every time I got to make one of those calls, and relieved when I no longer had to explain to teachers and other kids why it was we didn't have a phone.

I knew immediately when the caller was a telephone salesperson. If anyone asked to speak to "Mr. or Mrs. Walker" or "your mom and dad," that person didn't know my family.

One summer morning when I answered the phone a voice said: "Is this Gale's daughter? Is this Lou Ann?" I couldn't recognize the man's voice. He was speaking in a hoarse whisper. There was the sound of heavy machinery in the background.

The man began using words I'd never heard before, and the first few sentences caught me off guard. How it is we know sexual words we've never heard before is a mystery. The man on the phone was explicit and vulgar. I held the receiver tight. I didn't quite know what to do. And then I hung up and walked slowly back outside. Daddy was on the front lawn watering some plants.

"Telephone?" He signed with his thumb and little finger held to his ear and mouth as if his hand were a receiver.

"Yes."

"Who was it?"

"Wrong number."

But I was too quiet. He knew something else was wrong. I kept thinking it had to have been a man who worked with Dad and who knew him; the machinery sounded like the presses at the *Star-News*. I was scared the man would come get me and do what he'd said on the phone. A couple more times that day Dad questioned me about the phone call.

"No. It was a wrong number."

At first I felt guilty for keeping the secret from my father. Later I came to resent the man who had known for sure that the only one in our house who could answer the phone was a ten-year-old girl.

I got one more obscene call from him—or at least the beginning of one. Someone must have walked in on him, because he

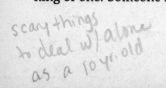

scary things to deal w/ alone as a 10 yr. old

hung up in the middle of the first sentence, just after he'd asked for me by name.

At school kids used to grumble about having to practice the piano. It was something they all seemed to have in common. I longed to be like the rest, complaining about études and scales.

"Mom, every kid has to learn to play the piano. Your sister Peggy plays. So does Grandma." I told her we needed music; we didn't have a radio or stereo.

My logic was irrefutable. A couple of months later, after great trouble and expense, a piano was delivered. It was a plain, refinished upright. The problem was, it was painted brown, not stained. The moment I saw it, I knew I'd never learn to play. I'd been envisioning a black baby grand.

The deliverymen put the upright in place and I sat down and began picking out a couple of tunes. Someone had given me a teach-yourself piano book and a few basic learner's guides and I hit the keys for "Merrily We Roll Along." Mom stood next to the piano, delighted. She kept moving her hand along the back, the front, the sides, everywhere she could, trying to find the best spot for feeling the vibrations during my one-note-at-a-time masterworks.

Ironically, after the expense of the piano, we couldn't afford lessons. For months I struggled, trying to teach myself a few songs. But I had so little natural ability that I never got very far.

And so the piano just sat there. Once in a while I would hear someone play a glissando passage, going through the whole set of keys. I'd rush to the living room, wondering who was there, only to find it had been Mom, dusting.

After several years I came home one day to find the furniture had been rearranged in the living room.

"Mom, what happened to the piano?"

"Sold it to some people. We put an ad in the newspaper. They had a daughter who wanted to learn how to play."

"But you didn't even tell me you were going to do that." I slapped the index-finger sign for "tell" from my mouth to my chest. I had no right to be so angry, but I was.

"No one plays it." There was no reprimand in her signing—she was just stating the facts.

Later, I took up the violin. I showed a little promise the first year, although the school orchestra leader became increasingly doubtful that the reason for my shortcomings as a stringed-instrument player was my inability to see the music stand. The stand came ever closer to the end of the violin—to the chagrin of my stand partner—until I was fitted with white, harlequin glasses. My playing, curiously enough, did not improve. I was reduced to years of violin mediocrity. The only good thing about it was that my parents never knew how bad I was.

Once a year the school orchestra held a concert. For weeks in advance, the teachers exhorted all the students to bring their parents and friends. At first I didn't want to ask Mom and Dad, but after the orchestra leader put on more pressure, I told them about the concert.

Dad fell asleep during the first selection. I could see him out of the corner of my eye. Not only was I mortified because his head was drooping to his chest, but I was scared to death he'd start snoring and Mom wouldn't know to stop him. Mom elbowed him awake twice and then gave up until the end of the first selection. A split second after she saw everyone else applauding, Mom hurriedly tapped Dad's arm so he could clap too.

At the end of the concert, Mom and Dad told me what a good job I'd done.

"I liked watching all the bows go back and forth together," Mom signed, mimicking a violinist drawing a bow. (Fortunately, even when I couldn't figure out which notes to play or where we were in the score, I was certain to keep my bow going in unison with everyone else's.)

When I finally decided to give up the violin in high school, everyone was relieved. As for Kay and Jan, Kay was first-chair viola and president of the orchestra all through high school, and Jan played excellent cello. But by the time their concerts came around, Mom and Dad had pretty much worn themselves out on the novelty of watching bows fly, and more often than not they begged off.

In a deaf household, there are certain things, quirky things, things that seem like the most natural happenings in the world, that a kid has to figure out alone. I vividly remember the first time I had a conversation with Kay when she was in a different room from me. Shouting back and forth, with a wall between us, in our own home, seemed surprising and unusual. It just couldn't be possible that you could talk through a wall—we'd always had to be in the same room with our parents in order to have conversations.

"Eenie, meenie, chili beanie, the spirits are about to speak!" Kay was quoting her favorite cartoon show through the plasterboard.

"Are they friendly spirits?" I shouted back.

"Hey, Jan, what's the next line?" Kay was calling through another wall to the back bedroom. Suddenly every corner of the house was alive and vibrant.

It never occurred to any of us to run away from home. I figure that's some kind of record for a family with three children. We hardly ever acted up. The most boisterous we got was when we chased each other round and round the circle path from our living room, through the kitchen, into the hall, and around again.

Running away from home is a stage most children have to go through, but it seems we were so protective we couldn't actually leave. Still, for a few minutes, we had to escape from our awesome adult roles. Instead of running away, we hid. (I have heard that a number of children who have deaf parents do precisely the same thing.)

Jan did it the most. Once, when she was about six, Mom, Kay, and I had searched the entire house, calling her name, looking in every closet and under every bed. She was nowhere to be found.

"Zhanli!" I whirled around. It was Kay doing a perfect imitation of my mother's voice. She was trying to trick Jan Lee. We looked at each other, both amazed at the reproduction of that eerie sound.

Mom panicked. But then she checked the closet and saw that Jan's coat was there. Jan had too much common sense to leave without it. Kay finally found Jan curled up beneath the headboard of my parents' bed, tucked so carefully that we hadn't seen her the first several times we'd searched there. Jan thought it was great sport.

Although I never realized it was happening, Mom and Dad made periodic rounds to be certain we were all right. They needed the reassurance because they couldn't hear what it was we were up to, and we couldn't call out if we were hurt.

My favorite escape hideaway was the back of our faded red Volkswagen Beetle. I would curl up fetus-like, devouring a book. I went through all of Sherlock Holmes in that tiny compartment. Cramped up there on a hot, sticky August day, I'd be transported to cool, foggy London.

I was in the middle of *The Hound of the Baskervilles* when I heard heavy footsteps walking past. (Many deaf people have a heavy tread because they can't hear the loudness of their walk. Some, like Mom, also have balance problems.) This time it was Dad. He called for me a couple of times; perversely, I decided not to reveal myself. I listened to him walk down to the end of the driveway and I knew he was searching up and down the street for me. As he came back up the drive, I scrunched down so that I wouldn't be seen. He walked around to the backyard, again calling for me. Dad didn't like to do much yelling; he worried about disturbing the neighbors and, because it didn't get much practice, his voice didn't hold out for long. By this point the crook of my knees was sweaty and my right arm had gone to sleep; I wasn't reading anymore.

He came back, walked into the house and in a few minutes came back out, walked down the drive, and this time when he called, there was anger in his voice. Finally, tiring of the game, I climbed over the back seat and out of the car. Dad was still looking up and down the street when I came up behind him and touched his shoulder. He wheeled around, pointed at me, an angry look on his face, and signed "Where?" both his hands out, palms up, moving slightly from side to side. I took him over and showed him the little space in the car where I'd crouched. The anger melted into amusement over my choice of location, but he added, "Next time don't make me look so long."

Just as the three of us had our physical spaces to escape to, so did Mom and Dad. I used to think of it as summer nights & never-never land.

On hot, sticky nights, after dinner, Mom and Dad would go back to the patio. Mom would light a couple of short candles scented to keep the mosquitoes away. As a kid I never had the patience to go out with them. There didn't seem to be anything happening. They'd sit in lawn chairs in the dark, one on each side of a small patio table, the orange candle burning in between but barely giving off light. When I went out to look at them, all I'd see was the glow of Mom's cigarette. If I stood there awhile, I'd be able to make out their forms, Mom with her legs curled up on the chair, Dad leaning back, hands folded in his lap. They'd each be staring out into space. All day long they strained their eyes, using them for watching signs, staying alert to everything, and at night they needed to rest them. This was their idea of peace: the quiet in front of their eyes.

Though it was dark, I knew what Mom's face looked like. Saturday afternoons, having done all her chores, she'd sit at the dining table, and there I could see her staring off into space. The only way to reach her was to tap on her shoulder. Sometimes I'd have to tap two or three times before she'd shake her head and return to me. If Kay or Jan or I told her something during one of those trances, she'd nod her head, but we knew we weren't getting through to her. Her mind was somewhere else.

Before retiring to the back porch, Mom and Dad might take a walk around the neighborhood. Once in a while they'd go for a bike ride before dusk. Mom's sense of balance had been slightly impaired by the meningitis, so she'd take my fat-tired Columbia. She didn't feel at ease on a bike, and although Dad put a rearview mirror on it for her, she worried about not hearing cars coming up behind her. Evening rides in the car, though, were something she looked forward to with girlish delight.

I remember being bored by drives, but Mom found them exciting. There was so much to see, she'd tell me. I preferred being home with a book. Her enthusiasm at Christmas was overwhelming. She wanted to look at the decorations on every house in every neighborhood in Indianapolis, and she never tired of driving to Monument Circle to see what the city billed as "the world's largest Christmas tree"—never mind that it was a cement spindle with lights attached from base to top.

At other times of the year, Mom would look at houses along the way to get ideas of how to improve her own home, what flowers to plant, where to put an awning or shutters. Even though we couldn't afford most of the home improvements, she loved looking and planning and dreaming. She took everything in. She could spot minor changes in a split second, and she often, excitedly, turned to me in the back seat, pointing something out so that I would be sure and see it.

On summer nights after the drive, when Mom and Dad were seated on the patio, I'd go out to ask them something. I'd stand in front of Mom and start signing. She'd take my wrists between her hands and move them into the candlelight. I remember the orange glow on her cheeks as she cocked her head, trying to see what I was saying. I'd run through it once and she'd squeeze up her face and shake her head "no," and move my hands closer to the flame. I'd try again. If she still couldn't see, she would move her hand as if flicking a light switch—the sign telling me to go turn on the back-porch light. I'd run in and flick on the yellow lamp, yellow so as not to attract moths. It cast an eerie pall over everything, so I'd hurriedly tell Mom whatever it was I had to say, then rush to turn the light off again and run back inside, leaving them to their dreams.

8

Telepathy

I remember one Christmas Eve so clear, so breathtakingly pure. Soft snow began falling that morning, a crystal white blanketing the yard, covering the maple tree in front and its twin next door. It fell and fell, and a few minutes after the mailman had trudged through to deliver the last of the Christmas cards, his tracks had magically disappeared. Dad was at work. Each of us—Kay, Jan, Mom, and I—was in a different part of the house making surprises, each excited, dreaming of the reactions of the rest.

In the kitchen, my mother created endless confections, brownies and cookies covered with silver beads so shiny they looked like jewels, pecan pies, pumpkin pies, sculpted breads, even a fruitcake with succulent red cherries. Only the grown-ups would eat the rum cakes and cheese balls rolled in pecans. But she'd made a favorite for each of us. There was no more room in the refrigerator. She stacked confections on top of the icebox, on counters in the garage, and covered in cool back rooms.

On the floor of the living room, Jan sat, the dimples in her cheeks showing her effort, designing an ornament.

In her bedroom, Kay with the round cheeks and the pixie

haircut that made her cheeks look even rounder, Kay who would soon start stretching to become the tallest of us, was wrapping gifts.

Each of us had a color, one that had been assigned to us as babies, one that we would keep through adulthood as ours. Kay's was pale pink, delicate and shy, the color of a blush on the cheeks. Jan's was red, bright and happy and alive. Mine was blue, to pick up the color of my eyes, my mother said.

That Christmas we each wore that color, that hue that brought out the colors of our skin, our eyes, our cheeks, that complemented our hair.

Jan was covered with more sequins than the bell she was making; the glue stuck to her fingers, dabs of it spotted her red shirt. Kay was calling for Jan to come hold her finger as an anchor to the ribbon bow she was making.

The windows clouded from the heat of the stove and the front-door window had smudges at each level where we'd rubbed away the fog from our own breathing to look at the yard.

It snowed all day and we created all day. We heard the soft whoosh of a car in the driveway at five and knew it was Dad. Mom felt the floor pound as the three of us rushed to the door, and she followed close behind.

Darkness came early and we were glad of it. We turned on the outside decorations, the lights Dad had strung in the evergreen bushes and around the edges of the roof. The icy tips of the maple branches glistened, reflecting the reds and greens and golds of our lamps.

And oh, the smells. Each dish, more tantalizing than the last, released sweet aromas through the whole house.

Finally, it was time for dinner. The lace tablecloth, the silver, the best dishes were laid. We'd even dressed up.

But first we went to the living room for one last look at the tree and the presents assembled underneath. Jan ran to turn off the lamps. Kay gasped at the beauty of the tinsel and tiny lights glowing in the darkened room, her enormous eyes growing even wider.

And then we sat down to dinner. Jan sped through grace. She could usually get through her set four-line poem in three seconds. Mom peeked to see if it was done, and as we raised our heads, she grinned, signing an "m" that mounted to an arc and glided into a "c," her fingers rounded: "Merry Christmas," a sign that conjured up a holly wreath with red berries tied in a velvet bow in my mind.

The candlelight flickered as we talked, highlighting a smudge of butter on a fingertip.

It was Christmas, a celebration holy for the lump it put in our throats, for the exquisite perfection of our happiness. We were home. Together. Warm. Safe.

Mom came back from the grocery store sobbing. I'd known something was wrong almost as soon as she pulled into the driveway. Instead of coming through the front door to greet us, as she usually did, she went around back to find Dad, who was building a toolshed.

"What is it?" Dad signed from the length of the yard, seeing her waving her arms.

"Come into the house. I don't want the neighbors to see me."

As I approached the kitchen from the hallway, I could hear her choked sobs. She'd cashed Dad's paycheck that afternoon and then gone to the supermarket. Someone had taken the wallet out of her purse when she'd turned to get a can from the shelf.

We lived from week to week. A month that had five Wednesdays in it was a godsend because, the way Mom and Dad calculated it, that meant there was an "extra" paycheck. They called it "money good." They figured the monthly bills—mortgage and car loan payments, electric and gas bills, and Dad's union dues—on a regular four-paycheck month. The little bonus helped us pay off a bank loan quicker, or allowed us to splurge on something new—say a couple of lawn chairs, or a winter coat. Mom is a careful spender; Dad is frugal. But there were plenty of weeks when we gritted our teeth until the following Wednesday. Mom might sheepishly come into our rooms on Saturday afternoon to ask if she could borrow enough for bread and milk from our piggy banks. When the next payday arrived, the first thing she and Dad did was return the money borrowed from us. More than once, such as the week the water heater and the refrigerator blew up, I thought we were in terrible trouble.

But on this quiet, sunny afternoon, Mom was beside herself. I knew how judiciously she shopped and how sharply aware she was of strangers. Standing in the kitchen, it was hard for me to watch her recount how it had been in the checkout lane. Her hands trembled as she signed and she had to keep stopping her words to brush the tears off her face.

The cashier had rung up almost everything—Mom watching the register closely to make sure the amounts punched in were the prices on the packages. And then she reached into her purse and didn't find her wallet. She rummaged through the bag. She looked on the floor, in the cart, among the groceries—the whole week's worth of food. She had to fight back the tears as she held out her handbag to show the cashier what had happened. She wasn't sure if the woman understood her. The people behind her in line grew impatient. The cashier didn't know what to do. Mom made a gesture as if patting the tops of the groceries—a head of lettuce, cans of peas, milk, carrots, hot dogs, a chicken—indicating to the woman to bag them and leave them there at the end of the counter. She pointed to herself, to the distance, to her wedding ring, and then the floor in front of her, indicating that she had to go get her husband. She raced back among the rows of the supermarket, hoping perhaps the wallet had dropped on the floor. It hadn't. Then she had come home to tell the story to Daddy.

"Are you angry?" She looked up at him.

"It wasn't your fault," he told her.

She was upset that the theft had happened in her own grocery. She worried about how tight money would be. An entire week's paycheck was gone. But far worse was the fact that she'd been publicly embarrassed. She'd inadvertently created a scene. And there is nothing in the world she hated more. She felt the eyes of the other shoppers on her, just as she'd felt people staring at her—the sign for "staring" is the fingertips of both hands almost poking into one's own face—in so many restaurants, so many streets, so many thousands of times during her life. That intense self-consciousness, that feeling that she'd done something ugly and wrong—when she'd only just been going about the normal course of affairs.

Before taking Mom back to the grocery, Dad washed and changed clothes. Neither of them could ever feel comfortable going out in public unless they were clean and neatly dressed. Mom fretted the ice cream would have melted by the time they got back to the market.

"Please, don't worry about that," Dad said, taking her arm and guiding her out the door. "It will be all right."

That day our lives seemed extraordinarily fragile.

In fifth grade I won the school spelling bee and the next step was getting ready for regionals. I'd never asked Mom and Dad for help with my homework, but now that the words were longer, preparing became difficult. The teacher urged me to have my parents quiz me on the lists in the official book. I don't think she comprehended the complicated logistics involved.

Mom was eager to help—until she realized that she couldn't pronounce any of the words. Few of the arcane terms on that list had exact sign language equivalents. I sat on the ottoman across from her chair and watched her trying to sound out the words to quiz me. With the first one I could puzzle together what she meant. The second one I couldn't figure out.

"Mom, that's not a word."

She looked into the book and tried to say something. "Zhoodikeyous," she repeated.

I came around and looked at the finger she had holding her place.

"No, 'joo-dish-as,' " I mouthed to her.

We went through the same exercise for a few more words. She was getting annoyed. She wanted to help me out, not get a lesson in pronunciation. "Oh, I don't know these big words! I'm stupid!"

And then she made a sign that sent a chill up my spine. It was a slang sign meaning "I'm deaf," but it's crudely done, made by putting the thumb in the ear and turning the rest of the hand downward—almost as if the hand is a donkey's ears.

"Mom! Stop it! Try something else. I have to practice."

So we tried having Mom sign the parts of words she knew and my relying on memory, having studied the pages so often, to know what the various endings were. That didn't work at all. Then Mom tried finger-spelling the words quickly—so quickly she hoped it would be a challenge when I spelled exactly the same thing back to her. (In theory, it might have worked. Just as we do not see each letter when we read words, a proficient signer sees the shape of the word rather than each letter when reading finger spelling. Otherwise, finger spelling could never be read so quickly.) But in this case, it was ludicrous. We stopped. She felt bad for letting me down. I just wished I hadn't put her through all that.

At any rate, for several successive Saturday mornings, Mom drove me to the competitions. I got all the way to the state finals, and each time I got a word right, I gave Mom, sitting in the darkened auditorium, a big smile so that she would know everything was all right. She knew, too, when at almost the very end of the contest, I messed up. There are two *n*'s in "mayonnaise."

"Don't worry. I'm proud of you." She knew how rotten I felt. Hugging me, patting my cheek, she said, "Next year you'll do better. Maybe I can help you study more. Do you want to go for ice cream?"

Of course I did. Mom was a great believer in food as a healing balm to the spirits.

Every summer in those days I spent at least two weeks in Montpelier, visiting Dad's brother Bill and his wife, Margaret. I relished those weeks; they were my vacation, my only regression into childhood. My room, at the head of the stairs, had carpeting and a thick, quilted bedspread. In the headboard of the bed, Aunt Margaret had installed books she thought I'd enjoy, one of them Uncle Bill's primer. Essays he'd written in grade school, tucked between the pages, would float onto the coverlet and I'd read of the time his mother had punished him by putting his blocks in the oven. Inside another was a picture of my uncle taken on board a ship in the South Pacific during his World War II navy days. A door next to the bed opened onto the attic, which contained trunks full of old clothes and tools and family documents.

We didn't have those kinds of mementos at my house. Grandma and Grandpa Wells hated clutter and got rid of anything that had lost its usefulness, and somehow my father's brothers and sisters were the recipients of most of the treasures on his side. The kinds of things I enjoyed looking at and handling in the storeroom were the artifacts my mother loved. She would look longingly at an antique mixing bowl and remember watching her own grandmother blending pie dough. She didn't recall any singing, or the click of a spoon against the pottery, or the muffled thump of the roller hitting the counter. But in her face I could see the intensity of the memories flooding back to her when I'd go home and describe what I'd looked at and handled in the attic.

Dinner every evening with Margaret and Bill was served on glass plates with real cloth napkins, and after dinner, Uncle Bill would take me for a chocolate-dipped ice cream cone at the Dari Delite, then we'd go back home, where Uncle Bill indulged in a few drinks and told me about the future. He promised me a bright-red convertible of my very own when I was in high school. He also promised me the moon and the stars.

Certainly I didn't feel the pinch of money worries when I stayed with them. Uncle Bill was always handing me twenty-dollar bills so I could treat myself at the five-and-dime.

Aunt Margaret, a wiry, pragmatic type, took me on all her errands, and I loved watching how she handled the banker and the man at the post office. There was never any uneasiness in the way these relatives conducted their affairs. Uncle Bill left me alone for hours in his office, me playing with the electric typewriter, picking out the words for a letter home on the crisp, engraved stationery. Best of all, there was serious work for me to do there. I helped move flowers and chairs, held doors, and fetched whatever was required. How ironic it was that I felt such pride doing this busy work. The importance of the actual adult transactions I handled for my parents paled in comparison.

Evenings, after the Dari Delite and the spinning out of Uncle Bill's promises, I'd sit in the living room at Aunt Margaret's feet, clean from my bath—she always let me dust myself with her Chanel No. 5 talc—while she slowly combed my hair.

"You know, all my life I've wanted long red hair. I think it's so special," she said quietly. Her dark-brown hair was unvaryingly cut short, with marcel waves.

"You know, we're so proud of you, Uncle Bill and I," she said, leaning close.

"Why?" I asked.

"Oh, don't you know? You work so hard. You make good grades. We just couldn't be prouder."

Thoughtfully stroking my hair, she'd lean forward, her mouth just behind my ear, and whisper to me, confide all sorts of things to me.

"You know, your uncle Bill and I, we tried and tried and tried for years to have a little girl just like you," she said so only I could hear.

At first, her words soothed me, but soon I felt my insides turning to jelly. There was something so seductive for a ten-year-old in what she was saying. I longed for their position, their lives, their ability to talk about anything and everything. I was shy but tried to appear forthright when I dealt with clerks and bankers and car salesmen on behalf of my parents. In their household, Uncle Bill did his own haggling over price. Aunt Margaret returned the milk herself when it was sour. I was so self-conscious about everything that concerned my family. Nothing was good enough, we weren't good enough: our clothes, our food, our silverware—none of it. What an ache, what a longing, what a maze! At that very same moment I was deeply ashamed of myself. I loved my parents fiercely. I would jump to their defense at any time. And yet I had these horrible, treacherous desires.

And after the two weeks, when I got back home, I was terrified Mom and Dad would figure out the secret.

Human beings are dismal failures when it comes to communicating. We've had it bred out of us, just as the Russian wolfhound has had its fine pointed nose and sleek, shiny coat developed from years of breeding and cross-breeding. The animal is swift but just a showpiece. I remember being so dazzled by the adults I met, by the sophistication they had, the witty conversation and the facile remarks. But so often the pith was missing, or hidden deep inside.

On the other hand, Mom and Dad had startling, almost telepathic abilities. They could piece together what was happening at an event, even when no one clued them in. They seemed to catch every nuance of movement, every blink. They noticed when a man and woman were not getting along, when no signs were evident to other people. Mom and Dad were actively participating without talking. They didn't have to contend with words as smoke screens. And there were times when I felt they were reading my soul. It was equally disquieting one day to discover that their empathy was a trait my sisters and I had either inherited or acquired.

One Sunday evening after we'd gone to visit Grandma and Grandpa Wells, Kay, Jan, Mom, Dad, and I were sitting at the kitchen table having a light supper, just toasted cheese sandwiches.

"Grandma's face," I signed, holding my hands to my cheeks, then pulling my hands gently downward. Perhaps she had a virus, maybe she was worried about something, but on that visit she looked ten years older than she had the month before.

I couldn't sign anymore. Suddenly, tears sprang to my eyes, then to Mom's, Kay's, and Jan's. My father looked on pensively. Grandma might not have looked different to anyone else; as it turns out, she did come down with the flu. And even though it was unsettling to be in tears over the prospect of aging and mortality, there was something both mysterious and reassuring about the communion among the five of us.

My Grandma Wells's meal repertoire wasn't large, but it was time tested and reliable. Talk over dinner was always the same: So-and-so is going to have a baby; Marlene and Jerry are moving; Connie bought a new car. Emphasis on factual reporting of the goings-on. Little analysis. The accounts were so similar they seemed timeless. After a spurt of talk, there'd be silence, only the sound of forks scraping plates and an occasional "Pass the butter, please." Then we'd plunge into how "Johnny couldn't get a license plate for his truck because the office closed early." I was sitting, poised on the edge of my chair, alert, waiting for something to happen, listening so hard my skin would prickle.

Sometimes Grandma would take my mother off to a corner to tell her a secret. Grandma used a whole variety of communication modes. She'd start a sentence by finger-spelling and mouthing a word, then she'd draw some letters on the palm of her hands, act out a little bit of what she wanted, maybe finger-spell another word, and then mouth another. The whole time, she had her head pulled back, peering up at my mother's face through the lower half of her bifocals.

At one of our Sunday dinners, when I was alone with my grandmother in the kitchen, she bent over to take the rolls, hot and glistening, out of the oven.

"I read about a cure for deafness," she said, measuring her words, her eyes still on the rolls.

That statement of hers filled me with skepticism and dread. From early on, I'd read about plenty of "miracle" treatments. But these "cures" were for less severe forms of deafness, mostly "conductive losses" that occurred later in life—deafness caused when one of the tiny bones in the ears is missing or damaged by disease, thus creating a break in the mechanical chain that brings sound to the auditory nerves. My parents' "sensory-neural loss" is far more complicated. The microscopic nerves and their attachment between the inner ear and the brain were destroyed. Medical science hasn't advanced as far as nerve reparation in the brain. Unfortunately, the articles never distinguished between types of deafness.

But my grandmother would get her hopes up. And I would panic. What if some extraordinary transplant *were* invented?

What if science were only able to restore hearing to one of my parents and not the other? What a cruel twist of fate that would be. I'd rarely seen a marriage between a deaf and a hearing person work out. Even if some treatment did give them back their hearing, their whole lives would be changed. They wouldn't have the same friends, the same clubs. It would alter every single relationship. Say they both had their hearing restored. What if, as hearing people, they weren't compatible? I didn't want to see their marriage ruined. With that chain of reasoning, I concluded our home would certainly be broken up. In later years I would meet many parents who had deaf children and dreamed of miracle cures. But I never met a single child who expressed such a wish for his or her parents.

I began losing my own hearing when I was thirteen. I didn't tell a soul about it. I was too terrified.

First my sisters' voices and the television volume became ever so slightly fuzzy. Then I was feeling the vibrations on my violin more than I was hearing the notes. From fuzzy I went to indistinct, from indistinct to thinking a glass bowl was over my head. I was sure the whole thing was a temporary aberration, perhaps psychosomatic, perhaps a hysterical reaction—one aunt was always warning me that reading too much would take its toll. Within a week the problem was affecting my schoolwork. After a few more days, when it was obvious I was getting worse, not better, I told Mom and Dad. They looked at me strangely. Mom took me to the doctor the next day.

In the waiting room, the receptionist took my name and told us to sit down. We'd been there many times before, but still the woman behind the desk stole glances at us. I signed low and small to Mom. In public we tried to be as inconspicuous as possible. Sitting in the waiting room, we tried to ignore the open looks of one man; we set up a wall around ourselves and carried on a conversation, or sometimes, if we were feeling too cowed, we didn't talk at all. We just laid our hands in our laps. This day we went on signing; only the nurse's call intruded on our private circle.

Inside his office, Dr. Marsh, a vague, disheveled man, looked in my ears, then got up and left the room. My worst fears were suddenly confirmed. At least I already knew sign language, I comforted myself. But I was already so protective of my sisters and my parents, so charged with the weight of grown-up responsibilities that I was fretting over just how the family would get along if I went completely deaf.

Dr. Marsh came back with a pan full of soapy water and a giant syringe.

"What are you going to do to me?"

"Ear wax," he grunted.

"What?"

He said it louder. I signed it to Mom. She'd been acting as if she weren't all that concerned about my hearing problem, but now she leaned forward, a worried look in her eyes, concentrating heavily on my hands, one eyebrow raised. With the diagnosis, she leaned back, obviously relieved and a little amused.

After a few soapy squirts—it sounded like Niagara Falls inside my head—my hearing was miraculously restored.

"Use Q-Tips with baby oil," Dr. Marsh said.

I'd watched Mom and Dad inserting cotton swabs into their ears, but I'd read so much about puncturing the tympanic membrane that I was scared to put anything in there. It was all right for Mom and Dad, I thought; they didn't have anything to lose. After my bout of deafness, I swabbed regularly.

On the outside, whoever it was doing the talking had a power over us. Our signs were small and timid, and our faces were almost immobile. But when we were home alone, the five of us were transformed and the signing was large and generous. We made faces. We teased each other.

When Dad came home from work, my sisters and I would run to meet him at the door and give him a kiss. (We couldn't have just shouted "Hi, Dad" from our bedrooms.) Often as not, Dad was in a playful mood. He'd throw his hat on one of our heads. He'd start to hand one of us the newspaper, then hide it behind his back and toss it up over his shoulder, grabbing it backhanded just as we were about to catch it. He'd point to a spot on the floor, and when Kay looked, he'd catch her nose. It was never prolonged teasing, and after he was done, he'd hand the paper to us and pat each of us on top of the head.

Mom, much shyer in public, was even more playful in private. If Dad fell asleep in his easy chair before dinner, she'd come in to impersonate his spasmodic snoring. She'd lampoon herself too, showing how she'd gotten up groggy that morning and tried to brush her teeth with Dad's shaving cream, her mouth puckered up from the taste.

In a house where they could make all the noise they wanted, Mom and Dad were especially quiet, partly because they don't know how much noise certain activities make. They don't want to disturb anyone. When Mom is in the kitchen with pots and pans and she's banged them together, she starts. The vibration of the bang gives a little shock to her fingers, so she tries to handle them gently. And of course she doesn't want to dent her pans. I never heard my parents smack their lips or sigh. Both of them pick up their dining chairs to replace them under the table rather than slide them across the floor, and to cut down on vibrations, Kay, Jan, and I were required to do the same.

Mom is so conscious of making noise that every once in a while, out of the clear blue sky, she apologizes when she hasn't made a sound.

"Oh, excuse me."

"What did you do?"

"I hiccuped," she signed, her right hand making a brief jump up her chest. She covered her mouth.

"I didn't hear anything."

"Oh."

One night after we'd all gone to bed, Jan and I ran into each other in the hallway while getting drinks of water.

"Shhh! Listen!" Jan said, grabbing my arm. "Do you hear it?"

From the back of the house came a low, soft hum and a few little smacks.

"They're Mom's love pats," Jan said. "I love it when she does that."

I knew those little smacks. They were the ones she might give us clear out of the blue when we were reaching for milk in the refrigerator. She'd come up behind me, put an arm around my shoulder, hug me tight to her, and smack my hip, all the while the motor in her throat issuing a gentle, cooing hum. It was the same feeling I had when I held our puppy or somebody's baby, a surge through my arms just wanting to hold and squeeze the thing I had so much affection for. Jan and I giggled, thinking of Mom, nestled in her warm bed, and her little taps.

"Do you think they sign into each other's hands when the lights go out?" Jan asked.

It seemed like such a lovely, intimate gesture. I wondered.

We all took after Dad, playing small tricks on each other. Quiet, doe-eyed Kay, for example, developed an echolalia designed to drive me crazy. Even when I shouted, "Kay! Stop!" she repeated the words in exactly the same tone of voice.

My revenge was to get her laughing with her mouth full at mealtime and see if I could get her in trouble with Mom. My plan usually backfired; it didn't take my parents long to figure out who the troublemaker was. Sometimes Kay started giggling spontaneously, and I would get in trouble for that too.

My favorite trick, though, was one I played on Mom and Dad. I would tap my toe under the table. I started with an adagio rhythm. If no one noticed that, I'd switch to andante. And if I was really feeling my oats, I'd try for syncopation. Dad would turn to me and sign "please," his flat hand rubbing his chest in a circular motion.

I knew I'd gone far enough when my father spoke out loud along with his signs for me to stop.

Humor at the dinner table was another thing. Dad loved telling what can only be classified as "deaf" jokes. There is really no adequate way to translate them. If we were with close family members, interpreting as Dad talked, the jokes he told got big laughs. But when it came to friends or relatives who didn't understand sign language, they usually looked as if they were waiting for the other shoe to drop.

One story was about a deaf man who was driving in the country, when safety bars were lowered across the road at a railroad crossing. The train passed, but the bars weren't raised. Finally, the man went to the stationmaster and wrote him a note: "Please but." That's the punch line. The joke is that the sign for "but" is the index fingers crossed and then opening up, just the way the bars protecting train tracks do.

In another story, a deaf man went to a bar and sat on a stool next to a hearing man he knew. The hearing man and the deaf man wrote notes back and forth to each other. When another man came into the bar and sat down at the counter, he joined in the conversation. Soon the first man had to leave. The other two continued their note writing. Another deaf man walked into the bar and thought it very funny that two hearing guys were sitting writing notes to each other as if they were deaf.

A third story has a macabre twist. A deaf man lived next

door to a known robber and one day decided to help himself to the spoils of the latest bank job. The robber, suspecting the deaf man, got an interpreter and went to the deaf man's house to confront him.

"I don't know anything about it," the deaf man told the interpreter, who relayed that to the robber.

The robber brandished a gun in the deaf man's face. "Okay, okay, I'll tell. Don't shoot me," the deaf man signed to the interpreter. "The money is stashed in the oak tree in the backyard!"

The interpreter turned to the robber and said aloud: "He signs he'll never tell. He says he'd rather be shot."

Our next Christmas witnessed a strange confluence of events. By then I was in junior high school. Mom had spent a long morning at the shopping mall, Dad obligingly tagging along, carrying bags until they were so full he couldn't manage. He sat down on a bench, boxes piled alongside him, patiently waiting until Mom finished. Mom came home tired, but hurriedly put everything away, immediately wrapping the presents she thought would be hardest to hide. The doorbell rang. (Actually, it flashed and brayed, being the doorbell it was.) Mom went to answer it. A man handed her a card that read: I AM A DEAF MUTE   HERE IS A CARD WITH THE MANUAL ALPHABET   I WOULD APPRECIATE CONTRIBUTIONS.

Mom looked at the card and then signed: "You're deaf?" her fingers shaped in a letter "d," lightly touching her ear, then her cheek near the mouth. The man looked at her, then fled.

"I don't like that. It's not nice for deaf people to beg," she said to me in obvious disgust, "but people who pretend to be deaf—that's worse!"

Whether he was faking or whether he'd run away because he was embarrassed about begging from another deaf person, we never knew. There's a strain of Calvinism in my mother that's strong. Indeed, it runs in most deaf people I've met. Both Mom and Dad were appalled to learn that the double income tax exemption blind people receive might be extended to the deaf.

"I don't want that," Dad said. He felt it was some kind of handout. "We don't need it."

When it was finally time for gift-giving, Kay, Jan, and I ran around the Christmas tree, reading name tags and piling gifts in front of each recipient. Kay brought in the dog and we helped her open her package; Jan brought in Sylvester, our black cat, and vainly tried to get him to claw the paper off his catnip. They were rituals that had gotten a little corny over the years, but they helped prolong the excitement. Each person took a turn opening a present.

After the gifts were all opened and we tried on clothes and picked up the heaps of ribbons and papers, my sisters and I each hung up a wool knee sock on the door of the front-hall closet.

This year Kay and Jan were too sleepy to come with me to the midnight candle-lighting service at church. A friend's family picked me up in the middle of a nasty sleet storm. We arrived at the church at eleven. Only a few electric lights were on. The church was packed. I always thought the Christmas Eve service was the best of the year, probably because it was mostly Christmas carols—"It Came Upon a Midnight Clear," "We Three Kings of Orient," "Joy to the World!"

The minister began his sermon:

"While you are warm and snug in your houses, while the visions of sugarplums are dancing in your heads, in the heads of your children, you must remember the less fortunate. Remember Mary and Joseph, wandering without a home. But as we remember those less blessed than we are"—his voice dropped to a near whisper—"sometimes we forget those we should try hardest to remember." Suddenly the voice rose: "The lame and the halt." It thundered: "The blind and the deaf!" I looked up from my hymnal. He went on about the "unfortunate blind, the silent deaf, the halt who would walk," intermingling all that with curses and miracles and stars in the east.

At the stroke of midnight, Christmas, he stopped.

The church went dark, only one altar candle flickered. The minister lit his candle from the flame, then the altar boys, robed in red and white, lit the wicks of long brass rods from his candle. And so the light was passed slowly throughout the church, each person lighting the candle held by the one next to him. The burning flames grew brighter, we stood, holding our candles aloft, softly singing "Silent Night, Holy Night." My eyes stung, from the beauty of the light, from the purity of the song, and from the hurt of the minister's words.

I stood wondering what my friend and her family were thinking. Had they picked me up because I was the unfortunate daughter of the unfortunate deaf? Had I been a Christian duty? At home we'd just had a wonderful, storybook Christmas: presents, laughter, kisses, hugs.

What was "wrong" with us? I sat in the pew thinking about what might be "wrong" with other people—the man with the bad knees who couldn't play softball, another man who spoke through a hole in his throat because of surgery, the old woman with the withered hand, the young woman whom multiple sclerosis had confined to a wheelchair. To most of them we seemed worse off. But we didn't feel that way.

I hid all those feelings away, though. I would never have admitted the embarrassment I felt or the fears or anything else to the family who drove me home on Christmas Eve. I hid all that away from them and from myself, and not even a spy could have divined what was there.

9

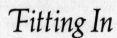

Fitting In

While other kids were going through adolescent rebellion, slamming doors, screaming at parents who didn't understand them, breaking all the rules, I was strangely silent.

In the late sixties and early seventies, the rest of the country was agitated. My neighborhood was as quiet and conservative as ever. The people who lived there had jobs at auto industry plants or at Western Electric. They worked hard. It was repetitive labor, but the hours were regular and their evenings were free for mowing the lawn and barbecuing hot dogs in the backyard, with the kids and dogs jumping around. The main excitement was "Hoosier Hysteria"—basketball—or preparations for the 500-Mile Race each May. That was it. People took a dim view if they thought you were unusual or you did unusual things. My meekness had one virtue: It kept me out of arguments.

I wanted to fit in. I was dying to fit in. I dressed carefully. I behaved carefully. And I wanted to do everything in my power to make Mom and Dad fit in too.

A banker whose children I baby-sat, a former neighbor of ours, was a member of the Masons. So were H.T. and Uncle Bill. When I asked the banker about the fraternal order, he told me there were secret rituals, that members had sworn not to divulge what went on, but he assured me it was a worthy cause.

I didn't understand what the Masons were, but I decided that if my relatives and this man were members, and they were all so successful and well liked, then my father should certainly join too. I don't know what I expected from it all for Dad, but somehow, mysteriously, it would help.

"But I won't be able to hear what's going on," Dad signed to me, truthfully. "I won't know what they're saying."

"No, you'll figure it out. They'll tell you," I urged.

"It's expensive." (Initiation fees and first-year membership were about five hundred dollars then.) "We need money for the family first," he told me.

"It's only that much in the beginning. Later on it's cheaper."

"You know I'm very busy already with meetings and other things."

"Yes, but Dad, this is important."

Dad got the application from the banker, and one day we drove to Montpelier to ask Uncle Bill for H.T.'s old lodge number and to get Bill to sign as a reference. When Dad handed him the form, Bill hesitated for a few seconds. I should have figured out what was wrong then, but I didn't. I assumed he was troubled because teenage girls weren't supposed to be involved in the application process, or something of that sort.

"Ah, well, yes, let's get this filled out," Bill said, frowning, searching through old files for the information.

Dad mailed the form. Several months went by without him mentioning a word about the application.

"Dad, what happened to the Masons? Haven't you heard from them?"

"Turned down." He used a "thumbs down" gesture.

I looked at him, stunned. "But why?"

He shrugged his shoulders. "I'm deaf."

He'd gotten the letter of refusal many weeks before, but had decided not to mention it.

My father's submitting that application had been a completely unselfish act. He had no desire to join a secret society. He didn't even know anyone who belonged to the local lodge. But Dad had always gone to tremendous lengths to please my sisters and me, the reasons for his selflessness bound up more in his gentle, courtly nature than in our complicated interdependence. True, I was willingly his interpreter, and he and Mom had to rely on me for so much. But they *wanted* to please us. We asked for very little. And they would do anything they could to give us the most normal childhoods possible. If it meant something as minor as joining a club, he would join.

As his thumb turned downward, I felt the blood rush to my face. I'd talked him into doing something he hadn't really wanted to do, something that wouldn't even have done him any good. He took the rejection well. That kind of thing had happened to him many times. But I'd never been the one responsible for hurting him.

It was all so complicated. We were father and daughter, yet often he'd had to defer to my judgment and let me be the one to control his transactions. I belonged to the hearing world, and in some ways that was the exclusive club he would never be good enough to join. I'd exposed him to this most recent taste of humiliation, and there was never the slightest rebuke from him. I felt as if I'd done my father a terrible wrong, exposing the differences between us that way. The episode brought me hard up against something I didn't want in my head: We weren't good enough.

The network of organizations Mom and Dad belonged to entailed every conceivable social, intellectual, and athletic function. Obviously, because the deaf population is limited, many of the clubs in Indianapolis had overlapping memberships, but still, the commitment of those members was amazing. Over the years, both Mom and Dad had served on dozens of committees. Mom was president of the women's auxiliary of one club. Together they were members of the National Fraternal Society of the Deaf, the Indiana Association of the Deaf, the Indiana State School for the Deaf Alumni Association, the National Association of the Deaf, the Indianapolis Deaf Club, and many others. At one time or another, Dad was president of most of these organizations. The clubs held dances, had basketball and softball leagues, dinners, endless parties and get-togethers. One or several of these groups sponsored nursing homes and children's camps, provided scholarships, crowned a Miss Deaf Indiana, and much, much more. One of the main functions of the "Frat" was providing insurance policies for deaf people, who were often denied coverage otherwise, or who were charged higher premiums solely because of their deafness. (There was no actuarial reason to charge more; indeed, deaf drivers have fewer accidents than the general population.)

If Mom and Dad had nothing else to do on a Saturday night, they might go to the deaf club to play cards or Ping-Pong, have a few drinks, then sit around chatting with their friends in sign.

Of course there were the usual number of arguments and upheavals. A man absconded with a club's treasury once. The political infighting was intense, and sometimes the elections seemed a little too heavily weighted in one person's favor. "That's the way things are everywhere," Dad once told me, shaking his head.

A few times Mom and Dad went to national conventions to represent Indiana organizations. Once they went to the Sherman House in Chicago. It was an old hotel with an antiquated fire alarm system. The management was concerned about having an entire contingent of deaf people. Upon registering, each conventioneer was given a whistle and instructed to pick up the phone and blow the whistle into it if there was an emergency. Dad became annoyed with one man, who, every time he wanted room service, whistled into the receiver. The third time the worried hotel manager, expecting a fire or a heart attack, raced up to the room, only to be asked for a whiskey and soda, he confiscated the man's whistle.

"Stupid," Dad signed to me, a fist knocking against his forehead. "Now what will that hotel think of deaf people?" That was what mattered to Dad. For him the conventions, where deaf people from all over the country gathered, were an important show of strength. But he was also concerned that outsiders not lump all deaf people together. *Stereotypes*

It was not long after the Masons episode. Dad and I were underneath the hood, bent over the car's engine. I was trying to listen to find out when he got an adjustment right—even though I wasn't quite sure what I was listening for. Uncle Garnel and Aunt Imogene had come to visit earlier that day.

"How come Garnel and Imogene never had kids?" I asked. I'd heard a whispered rumor from one of my cousins and I was checking it with Dad.

Dad looked at me pensively.

"Is it true that Grandpa Walker wouldn't let them?"

"Yes, that's what Garnel told me," Dad said.

"Why? I don't understand."

"I don't know," Dad signed. "It bothers me. I don't know if it's because he's deaf or what. But I think Garnel and Imogene should have had children. They would be more in the world. They would understand more things."

I asked Dad about H.T.'s imperious decree, but he was as mystified by it as I was. It didn't make any sense. Why didn't Garnel and Imogene just go ahead and have children anyway? And if it was the deafness, why had H.T. never ordered my father to do the same? I wondered if H.T. knew Dad wouldn't stand for it. I wondered if there wasn't something deeper, more mysterious. Someday I would find out.

In sign language, conversations like these are unbelievably hard. You must look directly at the person as you talk to him, and as he talks to you. You can't avert your eyes to relieve the tension. The contact is intimate and immediate at all times. And as you're expressing yourself, the ideas don't just go from your brain out your mouth. The emotion, the feelings circulate through your body, through the way you hold your shoulders, through your hips and legs and neck and cheeks and brows. The thoughts go through your head and arms and hands and fingers to someone's eyes. And although signs aren't a dramatization, there is such a close relationship between certain signs and what they represent that it can feel as if you're acting out much of what you say. You can't escape the emotion of a story. It reverberates through you. I flinched asking my father about the decree. My father flinched as he watched me ask, and as we talked, we could both see the ache. We were speaking in feelings. Words were not enough.

I had very few dates until junior year, when I met Dave Gregson, who was clean and neat—he washed his car before every date—and punctual, if not early. A few times as I madly pulled the hot rollers out of my hair, getting them all the more tangled, I could hear the strained conversation between my father and Dave as they struggled to pass the minutes. I knew what was going on, even though I could hear only a few grunts—Dave's—and my father's scratchy speech.

"Coke?" Dad said.

"Huh?"

"Drink Coke?" and I'm sure Dad leaned his head back in imitation of someone drinking a soda.

Dave would shake his head no, waving my father off with both hands. (That was another thing hearing people did—

made too-large movements. It was tantamount to shouting.)

Then Dad might pull the notepad and pen from his pocket and make an effort to discuss the weather or tell Dave what a nice car he had. Dave's responses were never longer than a word.

When I'd finally yanked all the curlers out and rushed to the living room, the air was thick. "Sorry," I said to Dave, signing at the same time to Dad.

"Jesus, why can't you be ready on time so I don't have to sit there like that? I hate it," he said, leading me to the car. He opened my door to make a good appearance for my parents.

"Sorry," I repeated.

One December afternoon—we'd been dating about six months—we were in the midst of a terrible argument. Dave was bulldozing me. I lashed out: "Why can't you be polite to my parents?" I was thinking of my mother's eager face when I left on a date, and of Dave's grunts. "You could at least try to make conversation."

"Jesus! What do you want me to do?" Dave yelled. "They can't talk, for God's sake. They don't have anything to goddamn say!"

And suddenly I realized that Dave, who was so concerned about new clothes and a clean car, would never want to be seen in public with people like my parents. Looking at us from his point of view, I saw how unusual my family was, how difficult we were to break into. I stewed about Dave and his reaction for quite a while. But what had happened to my aunt thirty years before had been far worse.

Gathel had not dated much until nurse's training, when she met Herb Engel, a medical student with blond hair and blue eyes, from a poor southern Indiana family. Herb's uncle, a doctor in Louisville, Kentucky, was sending him through medical school. During the time they dated, Herb had asked her lots of questions about her family, about her deaf brothers, and he'd spoken of marriage. At the beginning of February, during a long walk, Herb was talking about what a great team a doctor and a nurse would make, and then he told Gathel that when they got married she could never mention her two deaf brothers. She could not visit them and they would never be welcome at his house. There was to be no contact whatsoever. They had a terrible argument. Aunt Gathel told Herb she loved her brothers.

On Valentine's Day, when she got off the hospital's evening shift, she got a package from Herb. It was the hand-painted photograph of herself that Gathel had given him for Christmas. No note. And she never heard from him again.

That summer after graduation, my cousin Peggy, Aunt Gathel's daughter, was coming from Georgia for a visit. Peggy was beautiful, tall and leggy. She was my opposite: Where I was fair, she was tan. I had red hair and blue eyes; she had deep-brown eyes and hair, and a magnificent smile. And when she spoke, that soft southern accent was honey, the words easing slowly and gracefully from her mouth.

For years Dad's side of the family had been so embroiled in a feud, most probably spurred on by H.T.'s skulduggery, that hardly any of us saw the others. Only the coming of cousin Peggy could have brought so many Walkers together for a reunion. We decided on the fanciest restaurant in Indianapolis.

The day before the dinner Uncle Bill called, saying that one of my aunts had been speaking to him about the dinner. Dad was standing next to me. Receiver wedged between my chin and shoulder, I signed what Bill was saying.

"We can't wait to see you," Uncle Bill said. "But I just got this call from your aunt Diana and she thinks it would be better to go to another restaurant. She says the waiters won't be very nice to your mom and dad or to your aunt Imogene and uncle Garnel." I'd stopped interpreting for Dad at the beginning of that last sentence. I tried not to let my face change expression, and I jiggled the receiver as if my neck had a cramp and I couldn't sign. Dad's head was cocked and his forehead wrinkled as he tried to imagine what I was hearing. He reached out to help me ease the phone receiver, but I waved him away, then turned my back to him.

"Diana says she knows they wouldn't enjoy the restaurant all that much." Uncle Bill told me about alternate plans—dinner at the Holiday Inn on the outskirts of town. When I hung up the phone and Dad asked me what happened, I just couldn't tell him. "Uncle Bill says we have to switch dinner to another restaurant. I don't know why."

The idea that this aunt didn't want a maitre d' in a restaurant seeing her with us gnawed at me. I was so proud and sensitive. I hoped Dad wouldn't guess the reason for the change.

The next evening we all gathered and had a fine time. We laughed and joked and reminisced for hours. Aunt Diana and her husband never showed up. Nor did they ever phone.

At school I spent hours working on the newspaper and yearbook staffs, and I would spend hours during and after school talking with the journalism adviser. Spring of senior year, we had an all-school Easter convocation. The speaker was a local minister, and his talk was actually a sermon about the Resurrection and the nonbelievers he called "heathens." At this point I was going to church every Sunday morning, attending youth fellowship meetings Sunday evenings, and volunteering for all kinds of church missionary work in between. Yet I objected to having been forced to attend a religious Easter convocation at a public school. It seemed to be an infringement of constitutional rights. I did exactly what I felt a good journalist should do. I wrote an editorial citing the Supreme Court ruling on prayer in the schools. No fireworks. And I turned it in.

That afternoon, the journalism adviser told me the principal wanted me in his office. In thirteen years of schooling I'd never been to the principal's office, never been sent out to the hall, never had more than a one-sentence tongue-lashing from a teacher (from which I suffered greatly).

The principal's name was Obert Piety. He looked like Harry Truman, but without a bit of good-naturedness. He was spare, strict, and greatly feared. There were two assistant principals at Warren Central High School to see students who had committed infractions, the worst of which was usually letting a pig run down the halls during senior week. The most common problem was empty milk cartons flying across the lunchroom. Only high-level offenders were summoned to Mr. Piety's office.

"What is the meaning of this?" He shook my typewritten pages at me.

"That's my editorial. I don't believe students should be forced to attend a religious convocation. It has to do with the Bill of Rights. That was a sermon, not a public school meeting."

I was strong-willed for as much of the meeting as I could be. Mr. Piety told me that under no circumstances would a publication at his school print such heretical material.

"But it's an editorial. That means it's an opinion. What about the law? What about the Supreme Court?"

"There were no laws broken here. You should be ashamed of yourself, a nice girl like you."

That stung. He tore up my pages and insisted that I go back to the journalism teacher and apologize.

My voice quavered but I wouldn't let him see me cry.

"I didn't do anything wrong." And I didn't apologize.

The secretaries in the outer office stared as I walked out.

Most kids would have turned around and screamed. I couldn't. I could write an editorial about objective rights. I could argue for an abstract point. But when it came to defending myself, I couldn't. I felt foolish and nothing squashes resolution faster. Perhaps the principal was right. Perhaps I didn't have a legitimate complaint. Certainly the journalism adviser, who had lectured to us about the First Amendment, would not have gone to such lengths if I hadn't done a terrible thing. I stuffed every trace of rebellion back inside and tried to act as if nothing had happened.

I just kept wondering why my adviser had not come to me first with her objections to my editorial. Me, who had worshiped her. Me, who had only wanted to do what was right, what *she* had taught me. I felt betrayed. And later, puzzled. On Senior Honors Day, she gave me the award for best journalism student.

One evening not long after, I was out with my regular gang of girlfriends. I'd driven them to the Steak'n'Shake for burgers and fries, and as we pulled out, I squealed the tires. "That was an accident," I told my suddenly quieted passengers. "Honest."

"Know what my dad said today?" I went on. "He said he was going to take me to the doctor to have the lead removed from my foot. Pretty funny, huh?"

I don't know what came over me. Dad had said nothing of the kind. I'd made it up. It wasn't even that funny. This wasn't my way. I'm not a prevaricator. If anything, I usually downplayed stories rather than embellished them. It was just that I wanted to show my friends how witty my father was. The problem was that the funny things he really did say weren't all that translatable; they were mostly visual jokes, or Dad's teasing. But I'd felt compelled to make this comment up. When I quoted him to my friends I'd begun making him sound funnier and more intellectual than he really was. They couldn't possibly know that in his lexicon, my father really was funny and smart. I was just trying to show them that my dad was like any other dad.

We lived at the edge of suburbia—what seemed like the edge of civilization when I was in high school. And that night after I dropped off my friends, a curious ritual began. Five minutes from our house was flat farmland, stretching out to eternity. I'd head out a different road each evening after I'd dropped the others off. I wanted to see how many turns I could make, how far I could go without map or compass or any way of knowing where I was. I drove and drove and only allowed instinct to take me back home after twisting myself around on roads with numbers for names—700 W and 300 S. My sense of direction is terrible. I've never known west from south. I didn't dare look at the time or the odometer to see how far I'd gone. The few farmhouses I passed were darkened. The stories my high school friends had told me whirled through my head. Cowed by mothers who that morning had read about a man who stopped to help fix a flat and then killed or raped the woman driver, my friends would laugh nervously. Then they'd launch into a story about a hitchhiker who robbed someone and stole his car, or about a man who waited in back seats and underneath cars in parking lots. My head would be spinning with the sum of all those fears and as I drove faster and faster, the accelerator down to the floor, windows down, the breeze whipping my hair, the radio turned as loud as it would go, I felt free. The challenge was to see how far and how lost I could get, then wind my way out of the maze, trying not to backtrack. Some nights the tilled earth and the sky seemed one black splotch, a giant hungry hole all around me. Other nights I'd see menacing shadows in the six-foot-tall cornstalks.

There was a pounding inside me as I drove those roads, looking for something, wanting desperately to feel something.

Far from the tract homes and residential streets, away from the social climbing and petty jealousies of school, the gossiping and the backbiting, I was looking for a way out of having to be so dutiful. I guess it was sexual tension and middle-class tension and tension about trying to be perfect and mannered and polite that pushed me on those evenings. On the radio I played the bad-boy Stones, Mick Jagger, his husky voice shouting at me there beside the cornfields, "I can't get no satisfaction. . . ." Something in me desperately wished for more adventure, more of whatever it was shooting adrenaline into my blood-stream, pounding and pounding. Trying to get away. Yet aching to fit in. Trying to escape the ironies. Yet meeting reality head-on.

Learning

10

Broken Ears

Ball State University
Muncie, Indiana
1971

I had chosen a college geographically—and psychologi-
cally—between Indianapolis and Montpelier. The school,
named after the canning-jar family, had the only program in
the state for training teachers of the deaf, and although I had
not actively thought about it, that seemed like as good a career
as any.

The summer before college began, I'd done my first sign
language interpreting for strangers; I signed a computer course
for a young man who was learning Fortran. I didn't have a clue
as to what was going on, but I took my pay and the money relatives had given me as high school graduation gifts and purchased a typewriter, a dictionary, and a thesaurus. I glowed
with purpose and intellectual fervor.

I'd chosen to take a double English and French major along
with deaf education courses. Classwork was no problem.
Other things were.

"Well, *nice* Lou Ann, here is your mail." The girl behind the
dorm desk shoved a stained white envelope at me. I could tell

from the loopy handwriting that it was from my aunt Imogene. She'd addressed the envelope "To my Nice" (she meant niece), followed by my name. Imogene was married to Dad's eldest brother, Garnel. As with him, no one had ever diagnosed the cause of her deafness, although she, too, had been deaf from a very young age.

I waited until I got back to my room to open the letter:

Dear Nice Lou Ann,

Thanksgiving Day was over. We wanted to thank you send us a Thanksgiving Day card. very very happy to get your letter. I was thinking about going to write a letter and your birthday card this week. Garnel is a bowler every Friday. Just only for retirement. Yes he enjoyed to meet many friends there let him have a good time—what about me. Ha. Ha. I stayed at home and work some and read the Indianapolis Star every morning. The weather is nice and chilly. We will go to Ft. Wayne. Ind tomorrow (Sat) has a meeting Frat and small thing Christmas Exchange. . . . Have you a nice Visit your family for Thanksgiving Day? Most time we stay at home sometimes go to the Senior Citizen groups at Indpolis. Garnel has been running his nose and blow out—about a month. Well. Say Good bye—Hope hear from you.

Your Love—
Uncle Garnel & Aunt Imogene

always glad and enjoy to read your letter

The door flew open and my roommate, Cindy, strode in. Hurriedly, I hid Imogene's letter under some papers on my desk. I was too embarrassed to let her see it.

The truth was that Garnel and Imogene drove me crazy. Garnel made grunting noises when he signed—not at all uncommon among deaf people who can't hear themselves. Whenever he saw me, he'd pucker up his lips and grab my face to plant a wet, slobbery kiss right on my mouth. Imogene was always asking me to teach her to crochet. Patiently I'd go through all the steps, and then the next time I saw her, she'd ask me to teach her again. Over five years, she accomplished about four rows.

When we were out for dinner, Garnel pretended to look the other way when the bill came. At holidays my mother would search long and hard for the perfect presents for Garnel and Imogene, but my uncle refused to thank her for them, insisting that it was his brother who had given him the present. And my father didn't even get thanked if the present didn't thrill Garnel. On the other hand, Garnel was not exactly extravagant when he handed out gifts. Once he gave me a three-year-old fishing magazine.

I was self-conscious enough, but it was mortifying to be around them. My parents weren't at all like Garnel and Imogene. And I hated it when deaf people got lumped together, which is something even Dad's and Garnel's other siblings were occasionally guilty of doing.

One time Garnel and Imogene dropped in on me for a surprise visit at Ball State. I took them out for a snack, then a short walk around campus, ending at my dorm. I said goodbye to them at the fifth-floor lobby (the elevator stopped only on one and five), waving as the doors closed, then I rushed upstairs to finish a paper due the next morning. Entering my room, I heard an alarm ringing, but assumed it was a prank. (The university carefully segregated its male and female populations, and the fire door on the stairway between male and female floors was attached to a fire bell.)

As Garnel and Imogene were to tell it later, the elevator doors had opened, but they didn't think they were in the right place, so they rode back up to five, got out, and decided to walk downstairs. They couldn't understand why boys came pouring out of their rooms on every floor to look at them as they went down the staircase.

"Elevator—wait, wait, wait," Imogene signed later. "We

hurry home. Walk down stairs. Boys, boys everywhere, every floor staring at us. Why? We walk loud?" The alarm had indeed been loud enough to wake the dead. They hadn't paid any attention to the EMERGENCY ONLY sign on the door. Guiltily, I hoped nobody would trace them back to me.

There were several reasons why I had never told my father about the obscene phone calls back in the fourth grade. The first was for the same reason that so many child-molesting cases are never reported. It was my guilty secret. Like most children, I felt it was too bad even to admit. The second was probably, in my case, the more important. I knew there was nothing my father could do about it. If we reported the event to the police, I was the one who would have had to call them, or if we'd gone in person, I was the one who would have had to explain what had happened. I couldn't bring myself to do that. There was no way my father could innocently ask his hearing co-workers who it might have been. And it was impossible for me to describe what a voice sounded like to a father who had never heard one. It was a lesson in powerlessness. But somehow I had been able to shield my father from that reminder, and that was what made a difference to me. Years later when the same sort of vague feeling that something terribly wrong was happening—and again there was nothing my father could do about it—the grim lesson was held up for him and for me. I was embarrassed. But worse, I ached for him.

At Ball State we were required to take various physical education courses; I signed up for golf. Dad had played some tournaments at his newspaper a couple of times, and one weekend when I was home from summer school we went to a nearby driving range so I could practice. I was concentrating on all I'd been taught, trying out different grips. Dad answered questions if I asked, but otherwise didn't say much. He believed in letting me discover how to do things on my own.

At the same time we'd driven up to the range, I'd noticed a man in a yellow boat of a Cadillac pull into the parking lot. He was a heavyset man, flashily dressed, black hair slicked back. I was just hitting my third ball when the man came over and introduced himself so quickly I couldn't even make out his name. He said he was a golfing pro.

"You should be holding that club like this, see?"

"I'm fine, really. Thank you."

"No. Look here. You do it that way, you'll never get anywhere. Put your hands like this."

"I'm doing it the way I was taught at school. Thanks just the same."

He kept up a patter as rapid as a carny man's. "I don't like to see anybody doin' it wrong. Then you don't enjoy the game, you know? Here, sweetheart, let me show you how it should feel."

Dad had been standing down the row a way, practicing. Just then he looked up and saw the man talking to me. Dad walked over.

"He told me I swing wrong," I signed to Dad. "He says he's a golf pr—" The man's hand swept over mine just as I was about to finger-spell "o." Standing behind me, arms over my arms, hands squeezing mine, he forced me through a swing. This wasn't the friendly romantic gesture it was in the movies.

"Like this. It should feel like this."

He was pressing hard up against me from behind. I felt hot and uncomfortable. I smelled the thick oil of his skin. I must surely be mistaken about his intentions, I thought, trying to squirm away without seeming impolite. I just couldn't believe he would really be doing anything wrong—not so close to my home, not in public, certainly not in front of my father.

I finally got out from under him and swung once—a really rotten shot.

"No, now that's not what I showed you," and he tried to get behind me again.

"No. Leave me alone!" and I jerked away. Dad stepped forward, holding up his hand the way a policeman stops traffic.

"Let's go home," I signed.

We'd both been helpless in the face of this aggressor. As we drove home, Dad said very little. I think he was frustrated, even a little humiliated that he hadn't been able to shield his daughter from the unpleasantness. We never said another word about it.

I had never consciously intended to become a teacher of the deaf. My memories of the deaf school in Indianapolis were vague. I'd only been there a few times, when Mom and Dad went back for a basketball game or an alumni association meeting. It seemed a dark, forbidding place. I had the feeling that once someone got into teaching, there was nowhere to go. You stuck around until you retired. Yet part of me felt a sense of duty and obligation. No one was forcing me to take deaf education courses—unless it was those voices from childhood urging me to "do right" by my parents.

During my first course in the deaf ed department, the professor gushed over me. I was the only student she'd taught who'd actually known any deaf people. Most of the students in my courses were there because they'd seen a movie or read a Helen Keller book. For them there was something romantic and noble about the field. They were sure that with every child they encountered, they'd have a miraculous, spine-tingling breakthrough "w-a-t-e-r," just like little Helen's.

But I preferred their idealism to the worn-out, unimaginative nature of the professors. A couple of the teachers in the department would have nothing to do with me; they were wary of the fact that my parents were deaf. It took a number of years for me to figure out that I actually posed some sort of threat to them. At first I thought it was because the program at Ball State was "oral" (speaking and lip-reading to the exclusion of signing) and that my family used "total communication," that is, speaking in conjunction with signing. But that wasn't it. It had much more to do with an ideological need. None of these teachers had had any *real* success with children they'd taught. And one who had gone into deafness because her daughter was born deaf was heart broken over the fact that her daughter had rebelled, learned sign, then refused to have anything to do with her mother. What annoyed these professors was the persistent, nagging fact that none of their students had turned out more successful or happier than my parents, who'd been taught with "backward" methods. I was tangible proof of that.

One of my first courses was a practicum in teaching at the model preschool hearing-impaired program. Monday through Thursday, three- and four-year-old deaf children came to the morning classes. The children were loaded down with amplification, big boxes strapped to their small chests and giant headphones clamped over their ears. They looked like miniature spacemen. The teacher wore a microphone. It was the task of college interns like me to lead "positive reinforcement" experiments. Again and again the children learned to pronounce letters. The teacher held up a deflated red rubber ball. "This is a ball. Ball. Melissa, say 'ball.' " Puffing her cheeks out, the teacher led Melissa through "buh-buh-buh-buh—ball." There would be an M & M candy in it for the child who attempted the word (even if it sounded like "bog"). Early on the kids had grown savvy. They knew their lives would be markedly easier if they pretended ignorance for a while, and then suddenly learned "ball." They strung the teachers along at a prodigious rate for three- and four-year-olds. And they made a killing in M & M's. The teacher of that class was perfectly groomed, never a hair loose from her chignon, and when she talked, it was more like singing. She could smile even as the kids poured paint over her lap. But when she learned that my mother and father signed, the threatening tone in her voice was hardly veiled: "We don't sign here, of course. We want to prepare these children to enter the normal world." She believed that. And if she thought no one was looking, she would smack a child's hand if he so much as pointed at the "ball" without vocalizing.

I feel sure I would have fallen into Ball State's "oral" camp had it not been for Marion, the one deaf graduate student who started in the master's program that fall. Up until then, my knowledge about deafness was purely emotional, garnered from observing my parents and their friends. I knew nothing about statistics and philosophies of education. The program's teachings made good logical sense—it seemed right that people would learn better English grammar if they spoke it. It seemed true that in order for deaf people to become successful, they needed better verbal skills. After all, I wolfed down novels and I knew that my parents and their friends had a terrible, embarrassing time with even the simplest of sentence constructions. The deaf education professors' didactic views didn't disillusion me, nor was I analytical enough to figure out that their concepts failed to take into account the varying levels of innate ability within deaf children and the different times of onset of deafness. No, they confused me on a gut level: They treated that one deaf adult, a fellow professional, condescendingly—almost inhumanly. The professors rarely talked to Marion, even though she had had years of experience teaching the deaf—and being deaf. It was like the old joke: the operation was a success but the patient died. Theories and goals of education don't matter a whit if you don't consider your students to be human beings.

* * *

Sometimes the momentous decisions in our lives are ones we hardly make; they're so inevitable we simply tumble into them. I had not known anyone who had gone east to college. I was about to give up my home and a family that depended on me to go someplace far away. I was in my second year at Ball State, but I had nearly enough credits to graduate. Whatever it was that was pushing me on during those late-night drives that I still took through the darkened countryside, whatever it was that made me discontent, was sending me out to look for something. Yet I hardly knew I was serious.

"Where were you thinking of going?" my English professor, Dr. Rippy, asked me when I approached her. I told her.

"I never would have suggested that to you, but I think it's the natural thing for you to do. It's an excellent idea." She wrote gracious letters of recommendation for me.

I'd applied to four colleges for transfer: Smith, Mount Holyoke, Yale, and Harvard/Radcliffe. With trepidation, I chose the last.

The reaction of some of my old friends surprised me: "What's the matter? Indiana isn't good enough for you? You have to go off to Hah-vahd?"

That summer I worked as a reporter at the *Indianapolis News*, often writing obits or the readers' hot-line column, "Herman Hoglebogle." (I'd solve readers' sewer and weed crises and then write about the problem in the paper.) I had to be at the *News* at 6 A.M. every morning, and then five evenings a week, I'd drive sixty miles to Muncie to act as arts editor for the *Ball State Daily News*. Luckily, the job paid more than the gas cost. Besides, I liked the excitement of putting out a paper. That was where I met Steve Praeger, the sports editor.

In the basement of the old creaky house where the newspaper was housed, Steve and I worked side by side, pasting up our pages, waxing and then rolling on bits of paper, hoping that an *e* we'd inserted at the last minute would stick all the way to the printer. Steve, who had just finished college, was hanging around Muncie finishing a graduate course and figuring out where he was going next. He was affable, smart, and not at all like most of the fellows at Ball State. I thought his laid-back style was charismatic. His principal worry seemed to be whether he'd go bald by age twenty-three. We saw each other a lot that summer.

My only reservation about heading east to college was leaving him. We talked often over the phone and once I told my parents I was going to visit a friend and then I sneaked off to Steve's place in Champaign, Illinois.

There is one terrific advantage in having parents who can't answer the phone. They can't receive a call from the person I'm supposed to be visiting. I just hoped my sisters didn't piece things together if a slipup occurred.

That time with Steve was exciting. Even though they were fairly innocent meetings—I was still a "good" girl, after all—I came back refreshed and happy. I was hopelessly in love by then and realized, driving back to Indianapolis, that I might betray myself by seeming too vibrant. My parents were quick to pick up on my moods.

However, I had underestimated my father in another way. He had been keeping a record of gas consumption on the Mustang. A few hours after I arrived home, he said, with some amusement, that we must have done an awful lot of driving at my friend's house. "No, we didn't, Dad. Why?"

"Well, it's not a great distance, but the meter shows over three hundred miles."

"I guess we did drive some." I cleared my throat. I was convinced he'd be furious. He wasn't at all. Whether he guessed the truth, I've never known.

11

Commencement

Arriving at Harvard
September 1973

Everything in Boston was new to me. I'd never eaten yogurt or granola or tasted any kind of cheese but sliced processed. I was at once at home and uncomfortable in these surroundings. I dressed up nearly every day for classes. The uniform was blue jeans, but I didn't own a pair.

During that first month at school, there was a party at Adams House, the quintessence of Harvardiana, where I'd been told the undergrads had lockjaw accents and blue-blood New England family names. (I didn't know the Sedgwicks from the Saltonstalls.)

The party was noisy and packed when I arrived. Strolling over to the drinks table, I picked up a glass of orange juice. Right away, a tall, slightly scruffy blond fellow came over to me. I was flattered.

"What does your father do?" I nearly choked on my juice.

"He's a linotype operator at the *Indianapolis Star-News.*"

He turned on his heel and walked away. I stared at his back

as he made his way through the crowd. I should have given him some sort of withering, Jennifer Cavilleri–type response. It didn't matter that he might have done the same to anyone. He'd struck a nerve. I fled.

What overwhelmed me at Harvard was that everyone I met was articulate. Even the shy, studious science majors had a strong command of the language. In my classes, students seemed able to extemporize on topics that hadn't even been assigned. I could see how different my preparation had been from theirs. In school I'd memorized facts and formulae and could answer true/false and multiple-choice tests with ease, whereas these people had been taught to use the facts for the purpose of reasoning. They'd taken the numbers and dates and gone many steps further. In classes I watched as the others seemed to perform little dances at every meeting, pirouetting while I sat, a wallflower. It took a long time for me to figure out that many of the words I was hearing were decorations, extra flutes and hand motions to trick the ear into thinking there was more substance to answers than there actually was.

I decided I'd practice talking at mealtimes, when everyone seemed intent on conversation. These were real discussions. But I was so worried that my vocabulary was wrong or that there wasn't enough importance in what I was saying that I went over every word in my head before uttering it. By the end of my real narration, though, I'd added so many details and gotten so sidetracked by trying for perfection that my story lost its punch. I'd start out animatedly. "Listen to this!" I'd describe how a woman tried to stop me from getting on my bike by asking the time. "She'd been imported by some religious cult from Europe to be an undercover recruiter. And ..." I went on with the story, then panicked. I didn't have an ending line. I looked at the faces around the table. They were expecting something from me. I caught my breath. "Well, that's all ..." I murmured.

In sign, you more often than not start a story with the punch line. It's the telling that is important. You establish the basic facts of the story: "I saw two trucks collide today." But the trucks would be drawn in space and your two fists would collide violently. Then the story would head backward, explaining how you were driving down the road, looking at a lake, when suddenly one truck came rushing past you. . . . The fluidity of the sign is what the person enjoys watching, the actual telling of the anecdote, not the point it makes, not the final note. In sign you get excited about telling fairly mundane stories because the vigor of your presentation is part of the language. You're watching and feeling someone communicate. The language is so physical that signers are far more engaged with each other during a conversation than are most people who talk. You move and the other person moves with you. It's eyes and faces and hands and legs and torsos. Not disembodied words.

After the first few times people heard my descriptive vignettes and waited for the slam-bang end that wasn't there, I retrenched, not quite able to figure out why my stories were falling flat. I began to wonder whether it was because I was too easily impressed. These people knew so much more than I did. I went back to weighing my words carefully.

Thanksgiving came and another transfer student, Pat Ryan, invited me for dinner with her family, who lived nearby. That morning, before leaving, I made my first phone call home. Kay interpreted what I said to Mom and Dad as I talked, and she told me what they were signing. Occasionally she'd tell me to slow down, that I was talking too fast. At one point there was a pause and I heard some shuffling. I thought Jan was about to get on the line.

"Looahn." It was Mom's breathy voice. Kay had put the receiver to her mouth. Then I heard more shuffling. Mom had backed away as if scared, then looked questioningly at Kay. "Can she hear me?" Mom signed to Kay.

"Yes," Kay urged her. "Go on."

"Looahn," Mom said again. "I lahv you." She paused to breathe. "I mees you."

Then Dad. I could hear his preparatory swallow.

"My sweet daughter. I lahv you. Your Daddee," he said.

Kay came back on the line, but I was too choked up to talk to her or Jan. That was the first time Mom or Dad had ever talked to me on the phone. Their voices had the same distant quality of ships' horns on foggy nights.

Through Kay, Mom had asked again if I was coming home for Christmas. "Yes, of course I am. Tell her that, Kay."

For my senior honors thesis I chose a topic that had rarely been treated in literary essays: the confessional novel. At the time I didn't think the topic was at all autobiographical, but it's a little chilling to think how closely the themes of that thesis would parallel the themes of life during my twenties.

The thesis explored the reasons why protagonists of such novels as André Gide's *The Counterfeiters* and Ford Madox Ford's *The Good Soldier* had felt compelled to bare their souls. The subject was not so much confession as it was guilt and culpability, shame and repentance. The confessions themselves were so self-conscious that every word had to be examined and reexamined, and even then, one could never be sure that the final conclusion was correct. Everything could be twisted around to mean something different. In some of the accounts, the confession was a difficult, painful learning process. In others, the confession was an easy construct for absolving oneself of all guilt. The motive for some confessions was murder or cowardice. In others, it was an unspeakable evil of the mind, worse than any crime ever committed. There was no bottom, no end to the confession, just the knowledge that the chasm was ever deeper. These souls were contorted, but they realized that guilt was a black, bottomless pit and they could fall through it forever.

One evening when I was in the middle of doing the thesis, I was out with a new boyfriend. We'd been talking about those themes. Somehow the subject got around to my parents. Just as I felt I'd used my parents on my admission forms for college, it seemed to me I occasionally used their deafness as a kind of talisman. If someone asked me about myself, I invariably brought up deafness. I felt I was somehow lying if I didn't mention it; I would be distorting the truth by omission. After all, it was the thing that had affected every corner of my being.

But there was another reason I did it, a darker, far more complicated one. By giving someone the thirty-second prepared spiel, I could keep them at bay. It was like confessors in the novels for my thesis: If they admitted and were punished for a lesser crime, they were off the hook—in the public's eye—for any other wrongdoings. Their real punishment came with their inner torment, the agony they carried inside them.

For the first time I began telling someone about what I'd seen and heard, growing up. John, my boyfriend, was a sympathetic, reactive listener. It was the first time I'd ever trusted someone enough to admit that anything was less than perfect in my household. Tentatively I told him about being uninvited by our own relatives, about the golfing pro and the obscene phone calls, about children making pretend gestures and adults staring. With each story I cried. The litany "Be Good, Be Good, Be Good" was so loud in my ears, I'd never let myself cry before. But thinking back on all those times, I had this odd, inescapable feeling that society thought it was some kind of sin to be deaf. After all, something was making me feel terribly guilty. I had the urge to confess. But I didn't know what I'd done wrong. I knew that my parents hadn't done anything wrong.

My senior-thesis adviser, Joel Porte, was a kindly, professorial type, who couldn't understand my fascination with the convolutions of confession. Nor could he figure out why, halfway through, I suddenly became so uncomfortable with the subject. Every day, as I set to work analyzing those confessions, those crimes of the heart, the faces of my parents would flash into my head; my mother's delicate hands, my father's broad, handsome ones, moved silently through my mind. And my eyes would sting and my throat constrict. Frightened to go on, I was like a horse being led to the edge of a cliff. I was shying away. The topic had got too close. I didn't know what it was I had to confess. But on the other side of confession, I was certain, was the abyss.

I was in the cab on the way to the airport after commencement exercises. It should have been a day of celebration. Next to me was John, his hand over mine. Across my lap was a red rose he'd given me. I stared out the window on my side. The taxi driver kept looking in the mirror, waiting for the tears to come. I couldn't really speak. I was feeling too sorry for myself. I was about to fly back to Indiana.

The lofty words of the speaker were ringing in my ears, all those meant-to-be-inspiring words about how much awaited us in the world and how much we'd already accomplished. I was aching, thinking how much I would miss John. I was in love and convinced I'd never see him again.

The years had passed so quickly. I had studied and worked hard, made friends for life. I was learning to reason. And yet as I watched Boston go by for the last time, I realized that there was a great deal more I was supposed to have learned. I wasn't sure what it was, only that it was somewhere inside of me. But I knew I hadn't learned it. And going back to the middle of things I didn't understand meant I'd never figure them out.

12

Metaphor

Within four months, I'd moved to New York, found an apartment, and gotten an editorial job at *New York* magazine. Before I left home, Mom had said the same things she had when I'd left for Cambridge: "Can't you find a nice job in Indiana? Don't you want to stay home?" "Home" is signed with the curled hand kissing the cheek twice.

Dad held up a silencing hand. He knew I had to get out on my own. "You know. She must do this," he signed to Mom. "Remember? We did same thing."

"Will you come home and visit often?"

"Mom! Of course I will."

New York magazine seemed a very glamorous place to work. I was wide-eyed at the number of famous people I saw coming and going. The first piece I edited was by Margaret Mead. Parties were lavish. The editor's office contained Chesterfield sofas and a massive silver service. But by mid-December

147

something had gone wrong in the power structure and Clay Felker, editor in chief, was losing contractual control. The writers and editorial staff went out on strike in support of him and in protest to Rupert Murdoch's attempt to buy the magazine.

I went home for Christmas and told Mom and Dad that I was crazy about New York and my work. I had no idea whether or not I'd have a job when I returned, but there was no sense worrying my parents.

Just before they took me to the airport, Dad handed me a twenty-dollar bill.

"Here. For your taxi and things you'll need."

"No, Dad, I don't want it."

"Here, take it."

"No." I held my hands up to stop him. I just couldn't accept it. I knew Mom and Dad didn't have very much, but also I wanted to prove my independence. I saw a terrible, hurt look pass across my father's face.

Right after New Year's, following intricate financial and legal machinations, Rupert Murdoch took control of the magazine. Felker walked away with a million dollars. A few days later, my boss, Byron Dobell, was named editor of *Esquire* magazine and he asked me to go there with him. I'd been fortunate. The swirl of events hadn't left me stranded.

Although *Esquire* had been experiencing financial problems, Byron brought a new vitality to it and working there was genuinely fun. The only problem was that my tiny editorial salary didn't go very far in New York. I quickly saw that I needed to supplement my income, so I began a double life—teaching sign language and interpreting weekends and evenings while working at the magazine during the day. I figured I'd stop once my salary improved and I was writing more articles. But I didn't tell anyone at the office what I was doing. Once in a while an editor would ask me if I wanted tickets for a big movie screening and I'd have to turn him down in favor of spending the evening in a jail cell or a welfare office in Harlem. How could I explain?

I taught my first sign language class at the New York Society for the Deaf on East Fourteenth Street. Day or night, Fourteenth was creepy. People loitered. Stores selling housewares and cheap T-shirts out of cardboard boxes hired armed security guards. The nearby park was the exclusive province of drug dealers. The Society's building was not as bad as the rest of the neighborhood, but still shabby. It is a curious societal comment that the major agencies in New York serving the blind were on the genteel upper East Side or on tree-lined streets in Chelsea. The Society for the Deaf was in a place the police had forsaken.

As a child, one of the reasons I hated going to deaf social events was that they were held in seedy, decrepit buildings. The Indianapolis Deaf Club was above a porno peep show. The deaf community just didn't have enough money to support anything else.

I was apprehensive about going down to the New York Society at night and I was uncomfortable teaching sign. But I decided my students, all of whom were hearing, were really going to learn something. My lesson plans were marvels. If students asked me questions after class, I stayed and answered them at length. I even gave private tutoring sessions.

We were midway through the first term. I believed in total immersion, which was hard on the students, but I felt it would ultimately make them better signers. From the first two-hour class on, I had not said a word. I wanted them to know a little bit of what it is like to be deaf: lost, confused, unable to communicate. In the beginning I had to pantomime much of what I wanted to get across. I'd pretend I was turning on a set of faucets, holding out one hand like a cup. I pointed into the glass, then took it to my lips. "Water," I signed, the first three fingers of my hand pointing upright (a "w" shape), lightly touching the edge of my lower lip twice. If the students didn't understand the charade, I'd turn and write on the blackboard, but I tried to avoid even that. I didn't want the students automatically translating everything from sign to English—that slowed the process.

The class seemed to be catching on pretty well, except for one older man, who was having terrible trouble. His fingers were thick and slow-moving and he wasn't able to remember signs from one minute to the next, let alone from class to class.

I'd had the students fill out cards with their names, phone numbers, occupations, and reasons for taking the course. When I read through them, I discovered that this man was a psychiatrist at a large deaf school in New York City.

Because this was an "oral" school, all signing was forbidden; however, the psychological unit of the school was relaxing its anti-sign policies because the professionals felt that students who were having emotional difficulties needed more ways to express themselves. (Many of the children came from poor families, had often been neglected, and had virtually no verbal language.) I was pleased that this man was making the effort to learn how to sign, although there was a vagueness to him and he seemed unwilling to look anyone in the eye. I wondered how he had been communicating with his patients.

At the end of the fifth or sixth class, he came up to me as I was putting on my coat.

"By the way, what made you learn sign language?" he asked.

"My parents are deaf."

"Do you see them often?"

"No. They live in Indianapolis. I don't get home much."

As we were both passing through a narrow doorway, he said, not more than five inches from my face, "Children of defectives often feel guilt."

These words seared into my brain. He'd said the sentence as if it were the most natural thing in the world. Here was the man entrusted with the psychological well-being of young deaf children every day and he considered them "defective."

I hardly ever told my friends or the people with whom I worked about the interpreting I was doing. I didn't think they'd understand or approve. Besides, the sign language interpreter Code of Ethics required strict confidentiality and I took the code very seriously. As an interpreter (I was registered with the National Registry of Interpreters), I was never supposed to divulge any names or information about what transpired while I was working. Mostly I interpreted in courtrooms, jail cells, hospital emergency rooms, welfare offices, mental institutions, and college classrooms, occasionally signing for press conferences and television programs.

Some of the interpreting assignments were quite boring—classroom economics lectures three nights a week could be tedious. Others were upsetting. I had to tell one man he was dying of cancer. There were mundane assignments, filling out insurance forms or interpreting speeches at a banquet. The court work meant I had to do a lot of sitting around in filthy hallways, breathing stale smoke, and while I sat, I fretted about the chances I'd be mugged when I walked out onto the Grand Concourse in the Bronx. Often I had to fend off passes a court officer made. I felt I couldn't offend the officer (respect—or was it fear?—for authority was well ingrained), but also, I didn't want the officer getting mad at the deaf person. Or me. If he did, he could make us wait all the longer. So mostly I sat and stewed, thinking about my retort if deaf people weren't involved. Then I realized I wouldn't even be there.

More often than not, the doctor or judge or welfare agent assumed I was the deaf person's sister or neighbor or friend, even though I had carefully explained my role at the beginning of the meeting. According to the interpreter Code of Ethics, I was required to sign everything that was said, including a siren on the street or a cough in the back of the room. If disparaging remarks were made about the deaf person, I was to sign them exactly as they were made. My interpretation had to be faithful. If the deaf person was angry, then I had to sound angry. If it was the hearing person who was angry, then my signs would be hard, slashing through the air.

This work was throwing me into the intimate functioning of people's lives. From the vantage point of mediator between deaf and hearing people, I began to get an even clearer idea of what was separating the two worlds. Sometimes it was frightening to be in the middle of so many complicated transactions. Mostly it was frustrating. (I have, of course, changed all identifying details in describing the interpreting I did.)

The case of "El Mudo" troubled me for a long time. I just wasn't sure justice had been served. El Mudo (Spanish for "the deaf one") was a big, thick, affable boy from the South Bronx. It was his cheerful nature that had gotten him into trouble. He'd been driving the getaway car when three of his hearing friends—kids who were obviously using him as a lackey—had gone in to rob a dry cleaning store. The owner had balked, one kid shot him, and El Mudo had been apprehended as an accessory to murder. He was the first to go on trial. Two others had been arrested. The one who had pulled the trigger hadn't yet been found.

El Mudo's court-appointed lawyer spent no more than ten minutes with him in judge's chambers before he began plea bargaining. El Mudo said he didn't know the kid had a gun, or that they were going in to rob the store. He said the other boys had used rough gestures and told him to wait in the car. El Mudo had been pleased. They'd never let him drive before. He was only eighteen. He'd had very little education and his family was from Peru. His signing was virtually incomprehensible.

"He's playing dumb," the lawyer muttered under his breath.

"He knew." I wasn't so sure. I tend to be overly trusting, but then El Mudo seemed pretty trusting too.

It was summer. No one wanted a prolonged trial, particularly not El Mudo's lawyer, who kept saying he couldn't stand having to face the kid every day. "It'll drive me insane. This dummy will drive a jury crazy too."

The lawyer went out in the hall to talk to the family—at least ten people shuffled out after him in a clump. They were dressed like peasants and their eyes were filled with terror. Only one of them spoke English and he translated for the rest, telling them the lawyer was going to "cop a plea" for El Mudo. (I wondered how that translated into Spanish; I had to be careful not to translate it literally or El Mudo would really have been confused.)

The other two boys who had been apprehended glared at El Mudo when he was brought into the courtroom.

The judge also wanted a short trial, but he was honestly concerned that the kid get a fair hearing. The judge questioned me closely about deafness and the interpreting process. He wanted to know how much El Mudo was understanding. I told him as much as I was allowed to tell, but the Code of Ethics forbids evaluating a person's signing skills for hearing people, or making judgments about a deaf person's intellect.

The case began. El Mudo contradicted himself. He didn't understand the questions no matter how many times they were repeated. He kept changing the details. The lawyer took him aside to coach him, then approached the bench. "Look at how those two are menacing the dummy," the lawyer complained to the judge. "He knows he'll get his throat slit if he rats." The other two in the murder case had made bail. El Mudo had not.

El Mudo was sentenced to seven years in prison. He tried not to cry when the judge pronounced the sentence. The judge asked him if he had any comments he wished to make. El

Mudo signed that all he wanted to do was go home to his mother and father. "I'm good boy. I won't do bad," he signed. The bailiff allowed him to hug his family goodbye. The judge and all the lawyers knew that in prison El Mudo would be sodomized, abused, and beaten. In all probability, there wasn't a soul in the jail who could sign, let alone anyone who would try to talk to him. The innocent look would be gone forever. I went home wondering if I'd done enough, if El Mudo really had understood what was going on.

Not long after John Hinckley, Jr., shot President Reagan, a twenty-three-year-old deaf man was picked up for harassing a young Broadway actress. There were striking similarities between the two cases. Hinckley had written a series of love letters to actress Jodie Foster. Thomas Hansen, who bore an uncanny resemblance to Hinckley, carried a gun in the trunk of his car, and wrote letters to one of the stars of the musical *Annie*, pleading with the girl to return his love, warning her to stop drinking (he'd seen a newspaper photo of her next to a bottle of champagne, celebrating her eighteenth birthday), and informing her that he would commit suicide if she didn't permit him to visit. Hansen had been tracking the girl for six years—since the time she was twelve—following her across the United States. In court it was obvious he was from a middle-class family and it was also clear that there was something terribly wrong with him. He repeatedly refused psychiatric treatment, even though the judge, quite a perceptive man, told Hansen he could go home to his mother if only he would promise to enroll in a program to get help for his emotional problems. Otherwise, he'd end up in jail.

Hansen's mother pleaded on his behalf. Clearly she'd been pleading on his behalf for years. I'd seen the same thing happen all too often. The mother was overly protective of the son and was willing to get him out of any scrape because she felt so guilty about his handicap. Hansen's parents had divorced long ago; the father came to court but never said a word to his son, nor did he make eye contact. It was the father's gun that Hansen kept in his car while stalking the actress.

The judge questioned Hansen about his intentions toward the actress. "I love her. I want her to follow the Lord's teachings more closely." Each of these words I was saying aloud as Hansen signed them. Suddenly, Hansen became agitated. He told the court he did not want a fair-haired woman interpreting for him. He became abusive. "There are devils in her!" he signed. (It was indeed as if I were possessed; I had to utter each of his signs to the court in exactly the manner he was saying them.) "I want a man!" he demanded.

The judge told him he'd have to make do and purge his soul later. The mother pretended to look the other way during the outburst. Even though she was pleading for mercy for this son, I noticed that she didn't try to talk to him once the entire day.

There seemed to be so much pathos in every setting I was in. There was the case of the two young men being brought to court again and again for an entire year. Their accuser never once showed up. "What's the matter," they asked me. "Hearing people don't like us?"

Another appointment filled me with sadness and longing. I was accompanying a woman to an eye examination. She was a gentle deaf and blind woman in her mid-forties, who I thought was making a remarkable adjustment to the loss of her two major senses. But she was still haunted by the words of her mother, spoken years ago, a mother who'd warned her she'd never amount to anything. As we were leaving, the eye doctor said her ability to perceive light would soon be gone as well.

And then there was the court case of the young black woman whose sister's boyfriend had repeatedly handcuffed her to a pipe, had beaten and pistol-whipped her. The night he came in drunk and ate her baby's food, she pushed him away. He pulled a gun. In self-defense she stabbed him with a kitchen knife. In court she didn't know how a person was supposed to behave in front of a jury. She didn't know that society expected her to be hypocritically remorseful for the death of her tormentor. She didn't know that the jury would be appalled by the grunts she made trying to talk. She was found guilty. Just before sentencing, the woman's brother came up and asked me to interpret something to his sister. He told her that out of spite, the sister, who testified against her, had set fire to her baby's room and clothes.

In all these cases I was never allowed to interject a single solitary word. I was a robot. If I felt a psychiatrist was coming to the conclusion that the deaf person was a raving loon, when he or she just happened to be exhibiting a deaf mannerism that was completely normal, I could not say anything to explain deafness or its educational limits. I was powerless to straighten out the misunderstandings. And often the psychiatrist would be angry at me, feeling that perhaps I was misinterpreting what was being said. Often a doctor would order me not to sign something to the deaf patient. I had to reply that everything that was said I had to repeat in sign. This made doctors livid, even after I pointed out how unfair—and certainly rude—it is to tell secrets about people in front of them. It made me feel more and more helpless. The situation with Ray was an example.

I'd rarely had a client so overjoyed to see me as Ray was. He was sixteen, small for his age, part black, part Hispanic. He asked me—and everyone he saw, including the guards in jail—whether his mother was coming to get him. She'd been contacted and promptly moved away without leaving a forwarding address. Ray was charged with shooting a woman on the subway. I was to interpret his psychiatric evaluation.

Ray was filled with regret. A hundred times I must have interpreted his apologies to lawyers, psychologists, bailiffs, secretaries, everyone he passed. Ray had been riding the subway with a friend who had a gym bag. He and the friend had been chatting; the friend told him he had something special in the bag and Ray had playfully reached in and pulled out a cloth-covered object. The gun went off, Ray explained to the judge. Ray's lawyer had promised him she would get him out of jail, but there was a problem finding a home for young men that would accept him.

The first thing Ray said to the two prison psychiatrists was that he was cold. The two, looking and acting like Woody Allen parodies of Freudians, leaped on that remark, then began their Rorschach testing. One mumbled to the other: "There should be more sexual references. Don't you see anything sexual?"

Ray was befuddled by the tests on every level. He couldn't do simple addition and subtraction. One psychiatrist accused him of faking. The other asked Ray the date.

"I don't know."

In fact, Ray didn't know whether it was Thursday or Sunday. He couldn't identify the season as being winter, spring, summer, or fall. But finally the explanation for Ray's coldness came out. He'd been arrested in summer and it was now almost winter. All he had to wear was a thin shirt and pants.

In stating his home address, he gave a vague approximation of the spelling of the street, but he didn't have a clue as to what the number was. As the psychiatrist continued questioning him, Ray held his head and said he couldn't think. The session turned into a Gestapo interrogation. "Who was your father?" Ray, small-boned and smooth-skinned, shrunk in his chair. "I don't know." Apparently his mother had had a series of boyfriends, but Ray didn't know any of their names. Ray wasn't able to tell where he went to school, just that he had attended schools somewhere. The more questions the doctors asked, the more Ray's head hurt. He was terrified of them. He simply didn't know anything.

Ray was taken away and as I was getting ready to leave, one

psychiatrist said to the other: "Amazing. Definitely in the concrete stage. But how could he possibly have gone through school and not know the seasons of the year? Preposterous."

Ray's was an extreme case, but it wasn't preposterous and I felt incredibly frustrated with the two psychiatrists. They'd never dealt with a deaf person before in their lives, had no idea of what a toll deafness could take, particularly on a kid who had been ignored or mistreated his entire life. I cringed to think of what their report would say.

One of the criticisms leveled at deaf people is that they're rigid thinkers. For Ray everything was right/wrong, yes/no, good/bad. There is no room for gray areas in Ray's life. The reason why makes perfect sense.

When deaf children are small, everything is yes or no with their parents. It's easy to shake a head up and down or sideways. There are probably more nos, with slaps and shaking fingers, than there are yeses. If the parents and the children aren't talking, the children aren't picking up the filler between the black and the white.

When deaf children get to school, the struggle is always to get them to read and write English. They aren't encouraged to write fanciful stories, only "experience" stories—"Our Visit to the Factory," "The Rain," "What the Guinea Pig Did Today." The stories they read are Dick-and-Jane simple, and they read that sort of thing all the way through high school. In fact, an alarming percentage of deaf children graduate high school with a third-grade reading ability. Hansen's might have been that. Ray's certainly wasn't. But in all the drudgery of learning noun-verb agreement, gerunds, and participles, there's one thing that's never nurtured: fantasy. They don't get fairy tales because the teachers feel they're too hard to understand and impossible to explain. The teachers feel frantic—or disgusted—when there's so much to teach their deaf students.

It's little wonder that so many deaf people are at the concrete-thinking stage. Their entire lives are concrete, didactic efforts to do a-b-c-d. There are no intuitive leaps, no fairy tales, no dream analyses. Day after day in school, they get grammar pounded into them, and they learn to make "puh" sounds, and they try to make their Adam's apples bob with their *g*'s.

The most frustrating conversations I've had with deaf people have been about the gray areas. They get angry when there isn't a right answer to a question. Once I watched a boy in a vocational program put a motor together. The instructor was in a playful mood and began asking the eighteen-year-old, "What if I put this kind of pipe here?" The boy looked up and shrugged his shoulders. "What if this hose were connected a different way?" The boy just sat looking bewildered. With the third question, the boy got so mad he threw a hammer across the room.

And then these deaf kids get out into the world. They can perform the tasks they were taught, but any initiative was driven out of them long ago. The teachers wanted orderliness. There were few kids in a class, so they were watched carefully. The system works. The kids end up being exactly what the teachers wanted them to be: docile. They weren't training inventors or leaders or supervisors. They were training drones. Only some drones, hemmed in at every turn, end up throwing hammers.

For the rest, the whole thing works fine. I watched people my father's age, men and women who had pretty good jobs and relatives who lived nearby. They set up apartments when they got out of high school and they readily fell into a pattern. They worked year after year—never being promoted—but working steadily, rarely missing days. Then suddenly newspapers stopped using linotype machines or a factory closed and the person was out of a job. If the worker couldn't find another one, he or she ended up in a social worker's or a vocational rehabilitation counselor's office. I've heard plenty of job counselors mutter about how childish these deaf people were, how they didn't have any initiative. Yet no one had ever let them think that initiative was acceptable behavior.

Indeed, only a few research studies have been conducted on how deaf people think, and their findings are scanty. For the most part, they're correlated to the age at which the deaf person lost his or her hearing. What's interesting about the studies is that prelingually deaf people (people such as my parents, who were born deaf or who lost their hearing early on) have no interior monologue. In most of us there is a tiny voice we consult as part of our thought processes. Deaf people literally don't hear themselves thinking. Scientists speculate that we begin recording this inner voice from the second of our births, if not before. This is the voice that dictates all of our writing. A good writer, they presume, has a particularly well-developed voice. Many deaf people who have no voice sometimes have to finger-spell or sign to themselves to make sure of the spelling of a word or the phrasing of a sentence.

Research into the dreams of deaf people is also just beginning, but it has fascinating implications. Some researchers theorize that deaf people think the way others dream. That is, the deaf people's thought processes, because they are so pictographic, are really more like dreams. Interestingly, very few deaf people actually use sign language in their dreams.

It never occurred to me as a child that my parents' thought processes or dreams were any different from my own. They were my parents. I learned most of my behavior from them. How could they possibly think in a different way from me? And how did their lives and how did my life fit in with the chaos I saw around me?

I did go home for Christmas that year and it was as happy as any we'd had. Early that morning, my mother, my father, and

Kay, Jan, and I excitedly exchanged gifts, then we made the hour-long trip to Greencastle and my grandparents' house. Mom's younger sister, Peggy, and her family were already milling about the kitchen when we got there. Everybody was in good spirits. The only disappointment was that Grandma and Grandpa Wells had not put up a Christmas tree. Their trees were always scrubby—never taller than a couple of feet—but they were strung with wonderful lights, orange globes containing a magical, gurgling liquid. The absence of a tree was part of Grandma's effort to do away with the gift exchange. The one thing she didn't like was fuss—which was precisely what my mother adored, the excitement of celebration.

Grandma's one excess was home-cooked dinners. As we gathered around an overloaded table, she kept apologizing, saying, "Now, don't be too disappointed. Remember it's nothing special." Everything from the dumplings to the pies was good and hearty.

Aunt Peggy started up the conversation and as it flew around the table, Kay, Jan, and I took turns interpreting, one of us putting down a fork to translate, fingers and hands gliding through the air as my grandfather told a story. Mom and Dad, so hungry to know what was going on, hardly stole their eyes away as they reached for rolls or brought their own forks to their mouths. Mom started telling a joke, signing to me so that I would speak aloud—until she suddenly stopped and shook her hands as if erasing the air: She'd accidentally told the punch line.

After dinner we did the dishes and opened our packages, sneaking off to nibble pieces of fudge and red-sugar cookies. Midafternoon, feeling bloated and sluggish, Mom, Grandpa, my sister Kay, and I decided we needed to go for a drive.

We bundled up and went out to Grandpa's gray Chevrolet. Kay, now a college sophomore, her long hair tucked under a cap my grandmother had crocheted for her, took the wheel.

My grandfather sat next to her in the front seat as Mom and I climbed in the back. As we pulled out of the drive, Grandpa scooted around in his seat to face Mom and me.

"Tell Doris Jean I figure it's about time I learned that sign language," he told me. As he spoke, Mom watched my hands. Grandpa had never made a sign in his life.

"Never too old to learn," he said.

Gently we coached him. "Right"—first two fingers on the right hand cross, then the hand arcs to the right; "left"—thumb and forefinger spead in an "l" shape as the hand arcs to the left.

"Well, this isn't so hard," he said. Grandpa signed "right" and Kay turned the car. Then he made a "left" at the next corner and we went in the other direction. He led us into the old section of town, with its blind alleys and dead-end streets. Greencastle doesn't even have ten thousand people and Grandpa had lived there all his life. He couldn't help but know his way around. Still, he was concentrating so hard on his "left" and "right" signs that he had us turning into driveways of houses where heads would suddenly pop up in the front picture window, people wondering who was coming to call.

"I didn't mean that," Grandpa chuckled after the third driveway.

His signs looked like such gibberish that Mom started to laugh, then she leaned over the car seat to mold his fingers back into a "right."

"This vocabulary is none too easy," he muttered, staring at his hand. "My old fingers are so stiff and sore."

"This is the sign for 'car,' " and I started to show him what it was: two hands guiding a steering wheel.

"No, no," he said, brushing my imaginary wheel aside. "I'll just get confused. I want to get these down first."

We'd all been disappointed there wasn't any snow that

Christmas morning. Yet despite the day's bleakness, there was a nostalgic air in our car. Mom squeezed my arm when we came to the elm where her friend had had a treehouse and where they used to hold tea parties for their dolls when they were little. Mom made a gesture with her left hand as if she were balancing a tiny saucer near her chin, her right fingers pretending to hold a china teacup to her lips. The elm, black and naked, showed no traces of the little girls who had played there. When we passed the white clapboard house in which Mom had been born, she tapped Grandpa on the shoulder excitedly, motioning toward it. He couldn't quite figure out what she wanted.

"Mom says that's the house where she was born."

By the time we got back to his house, Grandpa couldn't reproduce the two signs he'd learned, but he hurried inside to describe to my grandmother how he'd tried.

The snow had suddenly begun falling fast and now Mom was in a hurry to head home to Indianapolis. Jan, still in high school, helped Dad load the car trunk with shopping bags full of presents. Grandma handed us desserts wrapped in aluminum foil to take with us, then hugged each of us, cake and foil getting smashed out of affection. As Mom began stepping out the door, Grandpa put his hand around her arm and pulled her back inside.

"Come here." He led her over to a corner by the kitchen cabinets.

"I want to tell you something." He stooped over toward her, his lips pursed out, the way he always talks to Mom, an exaggerated stage whisper.

"I can't remember if I ever told you I loved you," he said, deliberately pointing his finger at Mom with the beat of each syllable. She was staring intently at his mouth, her forehead wrinkled. "You're so special to me. I think about you an awful

163

lot and I'm proud of my wonderful daughter." He took a breath, then squeezed her hands between his and brought them up to his chest. "I love you very much."

Grandpa straightened up. Slightly embarrassed, he tugged at Mom's winter coat, pulling it tight around her neck.

She was smiling at him, her red-gold hair like a halo around her creamy cheeks. My grandfather was normally a plainspoken man. I felt I was eavesdropping.

Grandpa ushered us both out into the snowy night.

By the time Mom and I got into the car, Kay and Jan were settled in the back seat. My father had gone out earlier to warm up the car, but the weather was raw and we were still shivering, rubbing our hands together as we slammed the doors. Then we rolled down the fogged-over windows, foolishly letting all the heat escape so we could shove our arms out to wave goodbye. Grandma and Grandpa stood coatless on the front porch, blowing kisses.

We were rolling up the windows, still smiling, the yells dying out, when halfway up the street, Mom turned to me, puzzled. In sign language, she asked, "What was Grandpa saying in the kitchen?"

My heart froze.

In the dim light of streetlamps, I signed to her what I'd overheard. "Mom, he said he loves you."

In the country, leaving the lamps behind, it was too dark to see any more hand signs. Mom turned back around, clasping her hands in her lap. She sat with her head bent, contemplating something in those hands. I turned my face to the window, hoping she wouldn't turn around again and catch the glimmer of tears welling up in my eyes.

So much had been lost.

13

Vanilla Fires

Nature attaches an overwhelming importance to hearing. As unborns we hear before we can see. Even in deep comas, people often hear what is going on around them. For most of us, when we die, the sense of hearing is the last to leave the body.

I went back to New York haunted by that Christmas scene. I felt hurt, then angry. There was so much I didn't yet understand, so much I must have missed.

For years a friend had been urging me to write about what I knew—to write about deafness. After Christmas, when I saw I'd resolutely been avoiding thinking about what had been in front of me all those years, I decided the friend was right. Not long before, I'd begun hearing about a deaf street gang, the Nasty Homicides. Perhaps facing the issues I'd been shunning, perhaps setting down on paper what I knew about deafness, would help me figure things out.

It was not an easy article to do. I hung out in squalid tenements in the South Bronx; I waited alone on street corners where there was nothing but rubble for blocks around; I watched angry young men practice self-defense with brass knuckles and baseball bats. It was probably insanity on my part. I was defenseless in the most violent streets of the country. But the problems these deaf street kids faced reflected the problems of deaf people everywhere; focusing on these kids might do some good—for me, in particular.

I'd first heard about the gang through Pedro Acevedo, one of the leaders. Pedro fascinated me. He was deaf and his family came from Cuba. He spoke Spanish with his sisters and mother. His written English wasn't so hot, but his signing was extraordinarily clever. Pedro was a curiosity because he'd married a middle-class hearing woman who was a college professor. In his own way, he was probably the most effective deaf rights activist I'd ever met. At night he roamed the mean parts of Brooklyn and the Bronx with members of the gang. During the day he taught sign language, but he was always taking time out to get help for those he called his "brothers" in jail. Somehow he'd been able to lobby successfully for better treatment of deaf men in prison on Rikers Island. Pedro, small, squarish, bulldog-looking, convinced prison officials to put deaf men in cells together. In the past, guards preferred separating deaf inmates because they didn't know what the deaf men might be planning in their secret signing. Pedro argued that the deaf guys were already isolated enough.

When I first told Pedro I wanted to write about the gang, he was uncomfortable. He kept trying to examine my motives.

"I don't want gang members hurt. Too much bad stuff," he told me. "Police know too much already. Other gangs beat us up. Dangerous business." I thought he was being a little melodramatic, but then I didn't know much about street gangs. After several phone conversations, Pedro agreed to talk with the other leaders of the gang about my writing an article. Each section of the city had its own chieftain. Even this part hadn't been easy. Pedro didn't want people he worked with to know about his heavy involvement with the gang. I could only telephone him at certain hours when an interpreter he trusted was present. (I didn't yet have a telephone-teletype—TTY—for calling deaf people.) I'd talk, the interpreter would sign, then Pedro would answer and the process be reversed. Since these were really interpreter trainees, there were times when my message was misinterpreted and Pedro got mad. Other times, it seemed to take hours to complete what should have been a short, simple call.

Three months passed. Pedro had been busy, one leader hadn't shown up, another didn't want any part of it. Finally, Pedro and I met and he told me that Big Willie—whose name sign was made with fingers cupped like a large ear at the side of his head—had agreed to check me out. Pedro told me to meet him at 9 P.M. on a corner in Greenwich Village. He said he'd drive me to Big Willie's in the South Bronx. I took off all my makeup and jewelry and put on the most nondescript clothing I could find. I wore a scruffy down jacket and pulled a hat down low over my ears and headed for the street corner. I stood in the dark for hours, chopping ice on the pavement with my heel.

Pedro never showed. I tried calling him that night, but it was several days before I finally got hold of him again. He chose another street corner and told me to meet him there the following week. He didn't come that time either, or the next. I began wondering whether he was deliberately trying to make me look like a fool or whether he was testing me.

The third time his excuse had been that he wasn't able to get word to the Brooklyn leaders to come. The problem with a deaf gang is that when the group wants to get together for a rumble or a meeting, things are not easy to organize. The members can't call each other up, so runners have to relay messages throughout the five boroughs. Advance planning is required, and these were not exactly the kind of guys who kept social calendars.

The Nasty Homicides was actually an outgrowth of the

Crazy Homicides, a hearing gang that had alarmingly grown 25,000 strong in New York City during the seventies. Many of the deaf kids had been introduced to the gang through older brothers. By the eighties, the ranks of the Crazies had dwindled to almost nothing. However, a statistical bulge nearly twenty years before made the Nasties more viable than ever.

From 1963 through 1965, a nationwide epidemic of rubella—German measles—caused thousands of women to deliver babies who were handicapped in some way: deaf, blind, or retarded. Some had heart conditions or withered limbs. Many had more than one problem. Some eight thousand babies were born deaf, more than doubling the population of children their age in special school programs for handicapped people. Poor people living in the Northeast were particularly hard hit.

After months of waiting, Pedro was finally taking me to meet the gang. We were in the car, driving through areas more bombed out than Dresden after the war.

"You write good things," Pedro warned, sticking a finger in my face. "Gang must be strong. Not weak." Then he looked over at some guys hanging out on a sidewalk when we were stopped for a light. "Lock your door. Bad neighborhood." Pedro reached under the car seat and drew out a very long, thick knife.

The first meeting was in front of a grimy bodega. Someone had propped up seats pulled out of a car as a bench. There were five Nasties waiting for us there. They didn't look much different from the guys I'd interpreted for in the courts and jails. With my pale skin and hair, I felt conspicuous. I was on their turf. The Nasties I was to follow were predominantly Hispanic; the black deaf youths had a separate group.

None of the guys at the bodega that night had worn "colors"—the special gang uniforms. The Nasties had a skull and crossbones sewn on the back of their black jackets and they wore metal spikes on leather bands around their wrists and necks. They were checking me out more than I was investigating them.

"How you know sign language?"

"My parents are deaf." The Nasties seemed surprised.

"Why you want to write story?"

"I want to tell people about deafness. And I want to understand it myself."

"You'll tell the police bad things. You'll get us in trouble," Big Willie said.

"No, honest I won't. I'll tell the truth. I'll tell what I see."

Big Willie didn't exactly live up to his name. He was average height, thin but wiry. In part as a warning, Big Willie began telling me how hard it was to be a Nasty. "Never know when someone sneaks up on you," he said. "Deaf must have eyes in back of head."

Over the next few months a pattern developed. During the day I went to my white-collar job. (I hadn't really told anyone at work what I was doing.) On prearranged evenings, Pedro would pick me up and I'd hang out with the Nasties.

One day I happened to be talking to an editor at *People* magazine, Lanny Jones, and for some reason mentioned the gang. He immediately wanted the story and assigned a photographer. I told the gang about the picture taking. We went back to the testing. This time at least I had someone to wait with, Michael Abramson, the photographer. Luckily, I'd had the foresight to stipulate that we meet in bars, so we wouldn't freeze if we were stood up. Finally, Michael and Pedro met, though Michael wasn't allowed to take pictures for a while.

The Nasty Homicides ranged in age from about thirteen to thirty-five. They liked to boast that there were a thousand to fifteen hundred of them scattered around the city. I imagine it was closer to a hundred, although I never met more than twenty-five and usually hung out with about ten regulars. There were two reasons they exaggerated their claims. One was that they fancied themselves big shots. The other was that they wanted to appear strong and scare off the bullies in other gangs.

The Nasties had banded together for self-protection. Because they were deaf, kids had been picking on them all their lives. They liked to see themselves as vigilantes for deaf people. Their favorite sign was "Deaf Power," a fist over the heart and a fist in the air. These were kids who didn't fit in anywhere else. They never knew what the police and social workers were going to do. In fact, they'd developed a kind of paranoia, always having to look over their shoulders. Often they thought people were talking about them because things just happened to them, and it often seemed they didn't have any control over their lives. Someone was always losing a job or getting thrown out of an apartment.

Despite the fact that several of the guys had records, I felt there was something curiously naïve and beguiling about the Nasties. They were street kids who'd had tough lives, but they didn't seem to understand how to operate the way other street kids did. One day they wanted to go on a rumble. They'd planned it for a long time, but there was only one car. They decided to take the subway. They knew if they were caught with weapons on public transportation, they'd go to jail, so they talked Michael into putting their baseball bats and nonchukas (two sticks connected by a chain) in his car.

When they got to the rumble spot, Mendoza's Pool Parlor, there were only two guys there from the other deaf gang. Something had gone wrong with the rendezvous. The Nasties, decked out in their war costumes, paced up and down the street. I could see the outlines of people in tenement windows, looking out in terror as the gang marched in front of their building. Big Willie had shaved his head to appear more menacing. When he got worked up, his signs were very fast and he slammed his knees together nervously. Finally, it was clear that no one was going to show. Big Willie ordered his troops home.

I should have been able to see the humor in that situation, but I didn't. I was not at all afraid of Big Willie and his men, but it was the way the hearing kids in Mendoza's had eyed them, the way Mendoza himself had gruffly thrown Big Willie out when he'd gone in to ask someone a question, that troubled me. Despite their bravado, there was something so powerless about the Nasties.

Their ineffectuality increased my fascination. I was the one who had chosen to be there. I was the one most out of place. I was convinced that in the squalor of New York there were valuable lessons I wouldn't get anywhere else. As a journalist, I was there to observe and record. It was the first time in my life I didn't have to feel responsible for all the deaf people present. I was throwing myself on the streets to discover whatever it was I'd been sheltering myself from.

It was a strange kind of double life I was leading. At my office, things were going very well. I'd left *Esquire*, worked a stint at *Cosmopolitan*, and then gone to *Diversion*, a travel, arts, and leisure magazine for doctors. I had tremendous amounts of responsibility and the magazine had already sent me on trips to the South Pacific and North Africa. My office was demanding more and more of my attention, and I hadn't yet been able to make myself work on the article about the Nasties.

Still, I spent as much time with them as I could. Things weren't going well for them. One night, Juan came home from his after-school job as a janitor, to find his father blocking the doorway. In crude gestures, the father told him he'd sent all Juan's clothes to Puerto Rico. The father shoved his seventeen-year-old away and told him never to come back.

Miguel's parents had simply left for Puerto Rico. They didn't bother to write a note telling him where they'd gone—Miguel said they probably figured he couldn't read it. He and Juan were sleeping in abandoned buildings.

To me these were tragedies. The gang was fatalistic. So much so that hanging around with them, even in the midst of all this, took on a paralyzing sameness.

Afternoons the gang members filed through a graffiti-lined hallway to Big Willie's apartment. Big Willie's three small children, all hearing, wandered through the scene. Big Willie, José, and the others would pass joints and drink beer. Miguel would do push-ups, hoisting himself on just his thumbs. Noe kept talking about how he loved Arnold Schwarzenegger. José practiced Bruce Lee–type nonchuka techniques, expertly wrapping the two chain-connected wooden cylinders around his neck, his chest, his shoulders, and his waist. Big Willie's two-year-old son, Willie junior, was watching. José bent over to wrap the weapon around the child, trying to teach him the routine. But José was so gentle with the weapon that Willie junior looked up and smiled.

In the background a couple of stereos and a radio played rock and salsa full blast. A few of the Nasties had some hearing at certain frequencies. One wore a penny in his left ear to block out painful vibrations. They had mock fights, Noe especially. "Anybody say bad things"—he'd point to his nose—"I punch in the face. They bad to me . . ." Noe slowly drew his finger across his neck, indicating a slit throat.

One day we all went on a "peacekeeping" mission. By now I knew them so well I almost felt I was one of them. Certainly the fact that I spoke their language blurred the distinctions between them and me. Nonetheless, their lives seemed so hopeless. They were so easily thwarted at every turn.

Pedro had brought along Ortiz, a twenty-six-year-old who was working as a busboy. Ortiz was very thin, with a scraggly beard and mustache. He was missing one of his upper teeth. Indoors and out, he wore an orange stocking cap, which hid his hearing aid. It had been hard for us to talk in the car, and Pedro and Ortiz were hungry, so we stopped at a McDonald's for lunch. Pedro and Ortiz told me what they wanted and I relayed the order to the woman behind the counter.

We sat down and Pedro and Ortiz talked about fighting strategies, moving paper napkins and cups around the table, showing how the hearing gangs had surrounded the deaf kids' established territories. Ortiz said he wanted a rumble, and clasped his hands together to show a struggle.

Suddenly, Ortiz lost interest in the theoretical. He started talking about places he wanted to visit. It was cold outside. He said he wanted to see Florida and he pantomimed sticking one foot in the water at the beach.

It all seemed so unrealistic. One minute their pipe dreams were of bloody battles which they were too weak to carry out; the next, they were of beach walks. These young men couldn't read or write. They couldn't get or keep jobs. Many of them lived on government checks. And they weren't even all that good at being a gang. The social workers and police I talked to didn't want anything to do with them. They just kept stressing how hard it was to deal with deaf people, how you couldn't understand them, how they couldn't help themselves. The one hearing kid the gang looked up to was a drug dealer who handed out cocaine like jelly beans. Deafness was seeming like the worst thing in the world to me.

Ortiz borrowed my pen and in neat capital letters wrote on a napkin: "VANILA FIRES." He went to the hamburger counter and came back with a vanilla milk shake and French fries. His conversation with Pedro had been brutal—hands smacking hands, chests, foreheads. Still, his illiteracy had been serendipitously poetic.

When they weren't looking, I slipped the napkin into my pocket. The Nasty Homicides and I couldn't have been more different. But "vanilla fires" was an exact description of what was happening within me.

14

Quicksand

Nothing is as gloomy as February in New York City. I had hung around the ghetto long enough. I sat down to write the story about the Nasty Homicides. And got nowhere. The more attempts I made, the worse things came out. In one version they sounded weird; in another, wimpy; in yet another, vicious but stupid. I asked Anne, my roommate, to read a draft.

"It sounds as if you don't really want to explain deafness," she told me, "but unless you make yourself do it, people just aren't going to get it."

That wasn't the only problem. I was actually writing the gang story on the subway on my way to work in the mornings. I was in the middle of starting up a new national magazine, *Direct*, and everything that could possibly go wrong was doing so. And still, in the middle of the chaos, I felt compelled to slip out and interpret—evenings, weekends, lunch hours.

Every time I interpreted for someone, it was a different situation. Mostly I went home angry because the deaf person had been rude or the hearing people condescending. Sometimes I felt useless. But every time now, something pricked a memory of growing up. Each event conjured up the old conflicts about my parents and their deafness.

Once it was a college professor, a Yale graduate, who had hired me to translate a lecture he wanted to give his deaf teenage daughter. In long, scholarly sentences he told her that if she didn't buckle down and improve her grades, she would not get into college. And then what would become of her? I'd been impressed by the professor's calm manner and his concern for his child, until she excused herself to go to the bathroom and he turned to me and hissed, "That lazy little bitch isn't going to amount to anything." That was his own daughter he was talking about to me, a complete stranger.

Suddenly, I realized how things must have been for my own mother and father when they were growing up. I was being brought smack up against everything I'd put aside as a child.

I no longer needed the money, but I found myself taking on more and more interpreting. If I got a call from night court for an arraignment or if a hospital telephoned and asked me to help with an accident victim, I would cancel dinner with a friend to go. The pay was meager. It might take me an hour and a half to get somewhere, but I'd only be paid for actual interpreting hours. There were no benefits, no sick leave, no holiday pay. Something very powerful, some tremendous need, was making me drop everything to go.

One morning I went to a family court to interpret for a young man accused of attacking a woman with a knife while trying to steal her purse. He was only seventeen, and had been in and out of special programs for years. The problem with putting deaf kids in foster homes is that the foster parents rarely know sign, and the kid is just as lonely and frustrated in the foster home as he was in his own—if not more. This kid, Tyrone, was nearly six feet five. His face looked glum, but he was fidgety, the way kids are who've done angel dust.

We were sitting in a large room waiting for the case to be called. Usually before going into court I make small talk with the deaf person, in part to pass the time and be polite, but also so that I can get used to the way the person signs. Tyrone immediately began talking about his case. He told me he'd been falsely accused. I asked him not to tell me about it. "In the courtroom I'll interpret everything you say."

It's unethical to discuss a case beforehand. The client could get frustrated later and say, "I told you that before. Just tell them what I told you before." Also, I didn't want Tyrone to get worked up. He was making me nervous. But Tyrone wouldn't stop. He told me he was going to kill someone. Nearly every sign was "stab" . . . "kill" . . . "fight." He was getting more and more violent and I was the only outlet. His motions became wilder, exaggerated. He wasn't making sense; he was just working himself into a frenzy.

I went over to a guard and asked if Tyrone's case could be called soon. "Sit down and wait your turn," he answered. Patiently I explained that I was the interpreter and that there were other cases I needed to do in the next court. I didn't say anything about Tyrone's outbursts because I didn't want to get him in more trouble. "I'll see what I can do," the guard said. I sat down. Tyrone began acting out an entire scenario involving knives and guns. He stood up. "People die—I laugh," then he lunged at me with a stabbing motion. I didn't flinch. Calmly, I walked over to the guard. I didn't want Tyrone to know I was scared. I might encounter him again someday or I might have to leave the court building with him. On top of everything else, Tyrone seemed paranoid.

The case was called. The judge asked that the court appoint someone to do a psychological examination of Tyrone. The clerk handed him a card with an appointment written on it, and he and I rode down together in the elevator, him still telling me about how much he enjoyed killing. I knew Tyrone would be shuffled from court to court forever. He wouldn't really know why. In fact, his record indicated that he rarely showed up when he was scheduled for court, which got him into increasing amounts of trouble. I suspected it was because he couldn't read the appointment cards to understand when he was supposed to come. The only way Tyrone had of communicating was to hit someone.

Though I sympathized with Tyrone, I was terrified of him. I didn't want to be in the courtroom, but I couldn't refuse to go. The "vanilla fire" raged within me. While it burned on the inside, my facade was cool and sedate. The anger and confusion I felt watching these cases, seeing these reminders of my own growing up, racked my insides, but I couldn't let anyone know that. The appearance was so important. My words were careful, measured. It mattered what people thought of us. It mattered because ... I don't know why. It just did.

Back in my office the next day, I realized I'd promised a deaf man that as a special favor I'd interpret a meeting he'd scheduled with the head of his department at graduate school that afternoon. A Ph.D. candidate, he'd been having problems with two courses. The professor was threatening expulsion.

I had tons of work piled on my desk. And in the middle of everything, an interpreting coordinator called my office to ask if I would go to Brooklyn to translate.

"I'm swamped. Besides, I'm taking a late lunch to interpret someplace else."

"Oh, can't you change your appointment? A little girl has been raped in the Bronx and the D.A. wants to question her."

Immediately I began mental calculations, trying to figure if I could get to the Bronx and then back to the grad student's meeting, planning how late I would have to stay at the office to get my work done. And just as suddenly, a fury in me made me stop. I knew from experience that even if the coordinator told them I had to be somewhere else, the lawyer would arrive late or would be called away and I'd be stuck waiting for nothing. It would take as long as an hour and a half to get to the place and I'd be paid for two hours' work, even though I'd spend five. Worse, the coordinator wasn't even sure whether it was the little girl or her mother who was deaf. Many times people had played on my sympathies to get me somewhere and it turned out to be a false alarm.

How could I measure the Ph.D. candidate's future against the little girl's? And what about my job? I hung up the phone feeling terrible—furious, frustrated. All I wanted was to do the right thing—that compulsion that drove me on. And now all I felt was guilty and wrong. Manuscripts, photos, and typeset copy were flying around me. I was losing myself in the middle of everything.

The burdens. The should-haves. The issues were so complex, the misunderstandings so upsetting. Deafness seemed like quicksand. The more I did, the more I found there was to do. The more I did for one person, the more someone else wanted me to do. The more I found out, the more there was to learn. Everything became more, more, more. The more I saw, the more I felt guilty for not having done more as a child, for not speaking up, for not helping more.

I was the medium. I was the voice for what these deaf people said and I was their ears. I was a conduit for the gay arsonist, for the rapist, for the depressed teenager and the unwed pregnant mother. I was the one being yelled at by the psychotic father, by the doctor announcing cancer, by the minister damning the parishioner. And my hands were the ones the deaf-blind woman held as she was being told she'd lose even the tiny bit of light perception she now had.

"It's a magic thing you do," a woman from the audience told me after I'd stood and signed all the parts for a two-hour movie, my shoulders aching, my eyes blinded by the projector's light.

* * *

It was several days later. I was buzzing for entry to a hospital psychiatric ward. The attendant unlocked the door. I didn't want to be there that day. I was tired. My neck hurt and I'd pulled a muscle in my hand. A young, pretty deaf woman was waving at me. "I have an appointment with the psychiatrist," she signed. "Come."

She had a crippled knee and I followed her down two long hallways. It was a slow walk, the gait of a funeral procession.

"How are you feeling today, Miss Thomas?" the psychiatrist asked. I signed.

"I am feeling fine," Miss Thomas signed. I spoke.

"Let's talk more about your feelings of shame," he said.

"Yes, I was feeling ashamed because I am deaf," she said, matter-of-factly. "But not today."

Then she signed something else and I was sure I had misunderstood her.

"What? Sign that again, please," I asked.

"God came to my bed last night and sat down on the covers," she signed. "Jesus came with him. God told me I was Jesus' mother."

I talked. There was no emotion in my face, no other sound in my voice but precisely what was being said.

"I am the Virgin Mary." Suddenly the craziness of the whole situation hit me as I heard those words coming out of my mouth. *My* mouth. This was insanity. This was coming out of my mouth and it was madness. And I spoke it without emotion, without feeling. I spoke it. As I had spoken a thousand sentences before. "No, there is no difference between an orange and a banana." " 'Terminate' means someone who lives in an apartment."

"I am the Virgin Mary."

"Oh, and why do you think you are the Virgin Mary?" But this time the psychiatrist was addressing me. I pointed to Miss Thomas as I signed—reminding him who the patient was.

At least she was saying what she believed. I had been the medium for such insane words a thousand times in my life. And before that I had been the patient explainer to my parents and for my parents. I had explained and recounted and been the voice. A robot of words and sounds.

And suddenly I wanted to scream. For the first time in my life, I wanted to cry out. This isn't me. It's her! I have talked and listened and heard and there is no me! I have heard and hidden the insults so long, I have been the conduit so long that I am disappearing!

As I marched down the street from the hospital, my left hand was clenched. My right hand was restless. I found myself finger-spelling, swiftly, angrily: "I am *not* the Virgin Mary. I am *not* the Virgin Mary." Then I looked down at my fingers and shoved my hand to the bottom of my pocket.

My greatest fear had always been: The more you get to know me, the less there is to know. I was a black void. On the outside I was bright and shining and cheerful. Inside I was hollow. Deafness is also a void. It is the lack of something. Not the presence of anything. How could I be angry at what wasn't even there? Deafness had protected my parents. They were gentle souls in a harsh world, but there was a naïveté that shielded them. I'd thrown myself into that harsh world without any protective covering. And when I got bruised, it hurt. I could hear. I could speak. But I, too, was helpless. There was ultimately nothing I could do—or can do—for them. And so I was enslaved to do even more. I was a caretaker and I felt deep down inside I could never do enough for others because I could never make my mother and father whole. I didn't feel guilty because they couldn't hear and I could. It was much more complicated. I could never make them hear. And I could never make the world hear them. So what had I accomplished? What could I ever accomplish?

That night I had a pair of nightmares. In the first, I was protecting Mom and Dad from something. I'm covered with blood but they're all right. In the second, it's the reverse. They're bloody and I'm protected. I try so hard to scream, but I can't. Nothing will come. I wake from frustration. My mouth is dry but nothing comes from my voice.

15

All Through the Night

For a long time I was touchy as a sore tooth.

The next few months were awful. Consumed with fury, I snapped at everyone. I slammed doors. I cursed. It was as if I were going through adolescent rebellion.

I found I was sinking deeper and deeper into myself. I was struggling with an anger that got uglier by the day. On the outside I was trying to remain the "good" girl my aunts and uncles had warned me to be. The difference between the facade and the interior kept growing.

I felt guilty. It's easy to feel guilt. It's hard to absolve yourself. And never once had I ever been able to indulge myself in feeling bad. "You shouldn't feel sorry for yourself. Think of your mother and father. It wasn't *their* fault," I'd been told. In a bizarre psychological turn of events, I couldn't even feel sorry for myself without feeling the guilt. I had no right to feel bad, to feel lonely or out of touch or anything else. I had been boxed in on every side and that was what was struggling to get out. Just the right to feel.

I had been dutiful, obedient, quiet, shy, all the time I was growing up. And suddenly I was spewing forth. The venom was whirling inside me. I hurled vases to the floor to smash them. I slammed my fists into the walls. My fury came hurtling out at my boyfriend; the rage; the anger. I screamed at him and cursed at him. And when I'd poured it out over all those months, I was exhausted. I had screamed and spurted and spewed all the ugliness and filth festering inside me. Heaving like an animal after a bloody fight, I was finally, wearily, able to breathe. All those times when I'd carried the burden of deafness on my back. All those times when I'd been polite and said "Yes, sir" and "No, ma'am."

I got rid of my responsibilities.

I'd come close to suffocating under all those duties. I was nearly crushed by the unbearable weight, by the fact that if you have one responsibility, a hundred more come to rest on top of it: the responsibilities for my mother and my father and everyone else who had come along. I had stopped enjoying my life. I had become an automaton. And so I went to being a child for a while. I realized I'd have to reconstruct myself.

Sometimes when you hold yourself up to the light and scrutinize yourself mercilessly, there comes a release. I wish I could say mine was an overnight purge. But it couldn't have been. The wounds had festered too long. I had to learn I wasn't deaf. I had to start speaking out.

I'd known a few other sign language interpreters who had deaf parents. I ran into one woman at a meeting and found myself asking her about what it had been like in her household. I had cold chills running up my spine as I listened to her tell me stories that sounded exactly like mine. I told her about the time Mom had gone to pick up an ice cream cake she'd ordered for Jan's graduation party. The cake had the wrong wording on top. Mom was too embarrassed to have it changed. She didn't know quite how to explain it to the baker. Mom came home crying. Jan made a joke about the cake to the guests, trying to make Mom feel better.

The woman told me that her mother had once misunderstood the payment plan on a sofa. Seeing a sign in a furniture store window that said: $25 AND IT'S YOURS! she went in, paid the money, and had the sofa delivered. Then the mother started getting bills from the store. She hadn't known the twenty-five dollars was a down payment. The interest rates were prohibitive, and that was a time when twenty-five dollars was a fair amount of money. The mother fell behind on the payments. The store angrily reclaimed the couch. Her mother never again went to the street where the store was located. She still felt the shame of the repossession and the embarrassment of having misunderstood.

Hearing that daughter tell me of her own guilt and embarrassment and her intense identification with her mother gave me chills. It also gave me a peculiar relief. It was the first time another child of deaf parents had ever talked about the experience to me. I no longer felt so odd.

"Why didn't anyone ever talk to me about this before?" I asked the woman.

"You would have felt as if you were betraying your parents," she said.

It was too private. We didn't want to reveal the special, secret quality of our lives to anyone.

It was a long, slow spring. A spring against nature. I moped when I was alone and argued when I was with people. Although often shy, I was usually agreeable and wanting to please. But now I was on edge all the time. I quit my job in a huff. I had been pouring heart and soul into my work. It felt good to stop being responsible for all that staff.

And so I was putting my life back in order. I stopped seeing people I didn't want to see. I stopped doing things I didn't want to do. I tried to say what I meant. And I stopped interpreting. I had to halt that crazy addiction. The interpreting hit all the nerves of childhood. It was a daily replay of the hurts and the shame and the embarrassment. Only it was worse because now I had choices.

To earn my living, I was writing magazine articles and teaching, and I signed a contract for a children's book.

But there was still more to be reconciled. None of my relatives had ever actually talked about deafness and how it affected us—unless they mentioned the word and stiffened. Tentatively, I decided to broach the subject with my sister Kay.

Kay had recently been graduated from Northwestern University in Chicago and was now working as a vocational rehabilitation job coach, overseeing deaf clients who had come to her social service agency, unable to find work by themselves.

Carefully, I brought up the subject. "Do you think we're any different because of Mom and Dad's deafness?" I asked her.

"What do you mean?" She was wary.

"You know. You and Jan and I took on more responsibility than other kids. Don't you think that was hard on us?"

Neither of us had ever admitted it to another person, but it was true.

"I used to get so embarrassed in restaurants," Kay said. "Everybody would stare at us like we were freaks."

I told her that even now I had a hard time finishing a meal in front of other people. Eating made me too self-conscious.

"Sometimes it's so hard being different," I told her, my throat catching. "But I tried to act as if it didn't bother me."

"I felt so guilty, asking you for things, calling you up at college all the time," Kay whispered.

"Oh, no, Kay. No. Don't ever feel guilty. Please." I could hear the pleading in my voice. And that was when I realized that the sin was not within us. The guilt was. But not the sin.

Both of us began to cry. She and I had been so close all the time we were growing up. We'd talked about everything—except this.

I was finally able to finish the gang story. It was an unusual piece by *People* magazine standards. None of the characters could be construed as a celebrity. And it was long. The editors ran the article virtually intact—a rarity in a place where reporters report and staff members rewrite.

Taking Anne's advice, I tried to explain deafness, the frustrations, how difficult English was. The article was well received, but as I'd feared, there was an unfortunate consequence. Noe's parents, ardent church-goers, read the article and immediately shipped him off to live with friends in Puerto Rico. Both Ortiz and Miguel lost their jobs. Several of the other gang members had been forbidden to see their friends.

There was one thing that had always astounded me about the Nasties. Despite their talk about rumbles and protection, they didn't seem able to sustain genuine anger. They were living in squalid conditions with no hope for a stable future, and yet they were optimistic. Big Willie had put the pictures of himself in leather, spikes, and chains all over his wall. When I went to visit him, he told me about what had happened to the gang. He could see how upset the news made me.

"No, no, don't worry," Big Willie signed. "Those things happen. We knew before starting article what might happen. We enjoyed ourselves."

In other words, Big Willie was telling me that he had taken responsibility for himself. I had explained the consequences of the article in advance and he felt it was a true portrayal. And that was all he'd wanted.

I recognized that it would take a long while before I could become whole again. Now, instead of being angry, I was sad,

sad about what was lost and couldn't be regained. Sad about mistakes I'd made. There was a poignancy to everything. Each time someone asked me about my parents, tears came to my eyes. I had gotten a small TTY machine with a typewriter keyboard and coupler for telephoning them, yet even when I saw the electronic letters: HELLO GALE HERE GA (GA is typewritten shorthand for "go ahead"), I started to cry hard. I'd been sorting through memories and slights, crying and aching each time I put together a new realization. I'd lived my life with blinders on. I hadn't wanted to see how easily family mistreats family. Or human beings mistreat other human beings. The letters weren't enough. The electronic impulses for phone calls weren't enough. I decided it was time to go home and see my parents again.

I tried to act cheerful. My mother and her razor-sharp perception knew something was terribly wrong. She asked me what it was. I said, "Nothing," my fist starting at my chin, then thrown open. She didn't press. She simply said if I wanted to tell her, she wanted to know.

I leafed through old photo albums and asked her about her own childhood, about going to the deaf school, about her friends. She told me about dropping chocolate pudding in her milk, about a green dress she'd pointed out to her father in a store window. He'd bought the dress as a surprise. But instead of being happy with it, she felt bad because it had cost so much. She talked about little signs and gestures that stood out as symbols in her recollection.

Over the next few days I watched her. She must have felt as she had those times during childhood when I sat at her feet while she turned the pages of newspapers, me searching for spies. I was drawn into her day. Making a pie, she peeled an apple, the peel one long thin strip dangling off her knife. "Watch," she signed, and she threw the peel over her left shoulder, then whirled around to see what letter it had formed. "My girlfriends used to do this and the letter would be the letter of the name of the boy who liked us," she signed.

Another morning I started to peel a banana for breakfast. "Wait," she signed. "Think of a question you want answered—yes or no." I closed my eyes and thought. Then she took the banana and sliced the charcoal-black end off the bottom, and there was a fuzzy brown "Y." "Yes," she signed, smiling.

There is such optimism in my mother and father. I used to mistake their innocence for unsophistication. I used to be annoyed that they didn't know about the world. More recently I'd been angry that they hadn't fought back all those times they'd been mistreated. It was slowly dawning on me that innocence is a protection. They knew very well, better than I, how harsh the world is. And they realized early on they had one of two choices: to be bitter; or to enjoy what they had.

Dad and I were in a department store, walking past a display of electronic typewriters. Dad stopped at one. On the page above, people had picked out: "The quick brown fox . . ." Underneath, Dad typed: "I love my daughter Lou Ann."

In the past couple of years I'd noticed subtle changes in Mom and Dad. Since all three of their daughters had left home, they seemed to go out and enjoy themselves more. And there was a fortuitous confluence of events. Society on the whole, it seemed, was becoming a little more accepting. Less and less we were freaks. More and more we were curiosities.

One day we went to a Mexican restaurant. We were drinking slushy margaritas and eating messy tacos. In the next booth a man watched us, incessantly, pryingly, annoyingly, for the whole meal. We could all feel his eyes boring into us and we tried to ignore him and just eat our dinner.

As we were leaving, passing his table, my mother turned and

stared right back. I grabbed her arm, shocked. "Mom!" I fairly shouted in sign.

And then I started to laugh—at myself for my priggishness, and at Mom for her outlandishness, for her ability, finally, to stare back.

"I didn't like him doing that. It wasn't polite," she signed. "And so I thought he wouldn't like it either."

I thought about it for a second. "Good for you," I signed.

The next day Mom and I were on the patio getting ready for a picnic. I'd brought out a radio. She was shucking corn. An old Frank Sinatra song came on and I started humming along. I looked over at my mother and picked my hands up, signing: "Grab your coat and get your hat, leave your worries on the doorstep. Just direct your feet to the sunny side of the street"—"sunny" signed with the hand like a ball of fire throwing out its rays. "If I never had a cent, I'd be rich as Rockefeller. Gold dust at my feet . . ." Mom, hands full of corn, stopped, watching. Suddenly, feeling shy, I erased the air and started to pick up a platter.

"No. Do more," she signed. "I enjoy."

I finished that song and then a big, bright number came on. "New York, New York," I signed, and my signs became more and more expansive, my hips swaying, my shoulders moving in time with the music: "These little-town blues are melting away. I'm gonna make a brand-new start of it. New York! New York!" Mom looked almost transfixed as she watched me, and as the horns came to the climax, I extended my arms, my hands beating the air in time to the music, going for the grand finale. Mom applauded madly, laughing. "Wonderful!" she signed.

A few hours later we were in the car for an evening drive. She turned around to me in the back seat. "I never told anyone this," she began.

"When I was in school, in my room, when nobody else was around, I used to pretend I could sing." She was signing as if she were playing a guitar, her head thrown back, serenading us. "I liked to think I could sing." She laughed in that shy way of hers, chin tucked. "Nobody else knows that."

I don't think I had ever felt so close to her.

In June of 1983, Mom and Dad drove to New York City for a visit. This time I was really looking forward to spending the week with them.

The elevator in my loft building was broken and we had to climb the stairs. There was a terrible heat wave. The air conditioning in my apartment had broken too, so we ate out, trying to get comfortable. Only, in three different restaurants on three different days, the air conditioning was out of order.

We should have felt wretched, but we didn't. We were having a good time.

One day a teenage girl started staring at us as she was walking with a group of older tourists in the street. I knew Mom had seen her. The tourists were walking away, the girl trailing; her jaw actually dropped open as Mom and I continued signing to each other. In the old days I would have winced. This time I could shrug it off.

And then the girl walked smack into a pillar. Mom caught my eye and we laughed.

That evening we took a cab to dinner. Mom and I rode in the back seat. Dad was sitting next to the driver in the front.

"Nice weather we been having, huh?"

Dad didn't say anything—naturally. He was admiring the view out the passenger's side.

"Been here long?" Dad still didn't answer.

I was interpeting to Mom what the cabdriver was saying but couldn't sign to Dad because there was a bulletproof plastic partition between the front and back seats. Mom covered her mouth so the cabbie wouldn't know she was giggling.

For years Mom and Dad had been comfortable enough with

their deafness to make light of it. I always cringed. Mom thought it was particularly funny one time when Dad was painting the picnic table and Harpo, the dog, got her tail stuck to a tacky-wet leg. Her howls, at the one ear-splitting pitch Dad could perceive, made him spin around.

"You can hear! It's a miracle!" she teased him.

(Harpo was the same dog Mom had taught to sign. Upon signed command, Harpo could sit, stay, roll over, jump, and beg, among other dog tricks. The one thing Harpo rarely did was bark. She knew that with Mom and Dad around, it didn't do her much good.)

The next day I took Mom and Dad to the beach. The sun was setting and the sky was fiery red-purple. Mom and I stood, arms linked, watching the waves.

"What does it sound like?" Mom signed.

I wrinkled my forehead and held out my hand, palm up, hand searching, to show I was thinking of an answer. Then I made a little gurgling sound with my lips. Mom looked at my lips thoughtfully, then turned back to watch the waves hitting the rocks and made her own lips gurgle.

A minute later, Dad came over. He'd been watching some fishermen.

"What does it sound like?" he signed, pointing to the waves.

Mom grinned and signed that she had just asked me the same question, then she made the little lip movement for him. The three of us stood and watched the waves and the sunset. I tried to imagine the scene without the sound, and suddenly, everything before my eyes turned black and white, like a silent movie.

I took them on the Staten Island ferry. As we watched the cars being loaded and the dockhands at work, Mom grabbed my arm excitedly.

"Music!" she signed, and she moved to the beat.

"No, Mom! It's the engine," and I slapped a fist into my palm in time with the pistons.

"I think music," and she did a little dance.

I decided I'd drive back with Mom and Dad to Indiana to see my sisters. I had a few business errands to run, so the afternoon before we were to leave New York, I told them I had to go out and asked them what they wanted to do. Mom said she and Dad were tired and would stay in my apartment until I returned.

I was gone for several hours and Dad wasn't there when I got back.

"He went to see *New York Times*," Mom signed to me.

"But he doesn't know where it is. He doesn't know how to get there," I signed to her.

"He'll be fine," she assured me.

Dad came back half an hour later and told us about his day. Printers he worked with in Indianapolis had told him that the *Times* had a new laser process. Dad had wanted to see it first-hand, but we hadn't time. After I'd left, Dad had looked up the address in the phone book, and gone out to the bus stop near my house. (I'd taken Mom and Dad on a couple of buses but never this particular line.) The bus let him off in the middle of Times Square—which I always found intimidating—and he somehow made it to the newspaper building.

No one gets into the *Times* without an employee card or a pass. There are two imposing guards at the front doors. Dad went up to one man and wrote a note saying he wanted a tour of the building. The guard wrote back that public tours were conducted on Fridays only. He'd have to come back.

Dad wrote that he was leaving New York the next day and that he worked at the *Indianapolis Star-News* and very much wanted to see the composing room. No. He'd have to come back Friday. The guard busied himself with other people.

Signing the story to Mom and me, Dad stepped back and put his finger to his temple as if thinking, then he poked the air with it.

"I showed my union card," Dad signed, gesturing as if he were pulling out his wallet, "then I asked the guard if I could speak with the foreman of the local union. The guard telephoned him. The man came downstairs and took me all over and introduced me to other deaf boys. Very nice. Very interesting."

Dad then walked across town to Fifth Avenue, remembered that several years before I'd taken him on the number 5 bus, and came home.

The sixteen-hour drive back from New York to Indianapolis passed quickly. Mostly we talked. Every once in a while, when Dad noticed a peculiar town name on the interstate signs, I'd see him spelling the word in his hand, looking at his palm. Spelling "Wapakoneta" was his way of sounding out the pronunciation.

Two days after we got home, I asked Dad to take me to see the Indiana State School for the Deaf. He was surprised but pleased I'd asked him. Whenever Mom and Dad spoke of the school that had been their home for so many years, they talked of it fondly. Both of them were sad that the kids now were not as well behaved as they had been. But the place Mom and Dad talked about seemed to me to be in their dreams. The few times I'd been to the deaf school when I was growing up, it looked dreary and depressing.

Dad drove me over. He was president of the alumni association that year and he was going to see about arrangements for an upcoming picnic.

We went to the main office to get visitor passes. The secretary didn't know how to sign and never bothered to get up from her desk so that she could write a note to Dad. She couldn't read Dad's slow, distinct finger spelling either. She mistook him for someone else and made the name tag with the wrong name. Usually Dad would have let the whole thing slide, but this time he was on his own turf. He wanted the visit to be perfect. He asked that the name tag be changed.

The large administration building, the one my aunt Gathel remembered as long and dark, with a dangerous oiled floor, had recently been renovated. Dad had helped raise funds for that and several other improvements. This was the first time I was seeing all the work he'd done, work he'd hardly ever mentioned at home. We ran into a few teachers who had been classmates of Dad's. They spoke fondly of the old days and told me again how much of a sport my dad was and how much they loved teasing him. Dad's high school math teacher, also a deaf man, saw us in the hall and motioned for Dad to come in and talk to the class about who he was and what he'd done.

Dad continued the tour. All over the campus, noble inscriptions had been carved over the portals: "Truth is Beauty." "Seek the truth and ye shall find it."

"This building will be torn down soon," Dad signed as we stood on the grass. It was the dormitory where he'd lived for nearly twelve years.

"That one too," he said, pointing to the building where Mom had lived.

It had been a bittersweet day for both of us, but I was pleased I'd come. For so long I'd been doing what I'd accused other people of doing—I was seeing the deafness, not the people.

Mark Twain once described growing up very succinctly: "When I was a boy of fourteen, my father was so ignorant I could hardly stand to have the old man around. But when I got to be twenty-one, I was astonished at how much he had learned in seven years." My father hadn't started learning until I was twenty-one, and he didn't learn much until I was thirty. I realized I was doing what every child must do. I was rediscovering my parents.

16

Home Again

August 1983

Jan telephoned. I was sure she was calling to find out when I
was arriving home in Indiana for her wedding the following
week. I wish she had been.

A few days before, she'd called to say that Grandpa Wells
was in the hospital with a spot on his lung. Grandpa had always been so healthy and uncomplaining that I was certain it
was a doctor's scare to make him give up smoking. It was not.
He had inoperable lung cancer.

I told Jan I would get home as quickly as possible.

I never saw my grandfather alive again. He died the night I
flew to Indianapolis, never awakening from the coma he'd
slipped into. It all happened so quickly. The next morning we
drove to Greencastle and went straight to my grandmother's
house. Except for a couple of stints in the hospital, my grandparents had never been apart in the fifty-six years they'd been
married. They admitted to having fought plenty and they'd
both learned over the years to contain impatient natures, but
by the time I knew them, they'd either worn or smoothed each

other out. They took such pleasure in going out for breakfast each morning at The Point, a Greencastle diner where there was a crew of breakfast regulars. If they didn't go to The Point, they went fishing or mushroom hunting in the woods. They planned the rest of their lives around seeing the family. Their first great-grandchild was just three months old. "Oh, we're pretty happy," I'd hear my grandmother sigh from time to time. "We just take things as they come."

When I walked through her door that day, my grandmother burst into tears.

"He wanted it this way," she said. She was dressed in a simple navy suit she had made herself. Grandpa had helped her pick out the fabric.

"He wanted it quick." She'd been up most of the night, had bathed and dressed herself at 5 A.M. While we were waiting for the rest of the family to gather, she sat in Grandpa's favorite chair; I sat in hers. She picked at her handkerchief.

That day was filled with details: phone calls to friends and relatives who lived far away, arrangements for pallbearers, flowers, a minister. Mom and Dad were alert, watching everything. Jan and I took turns interpreting whatever it was an aunt or a cousin said. Grandma was too distraught even to attempt her semi-signing that day. Kay, who had recently married and moved to Lexington, Kentucky, was to arrive the following day. If Jan or I was called aside to do an errand, I'd see Mom out of the corner of my eye, looking around, wanting to give comfort and be comforted. She would go over and hold her mother's hand, gently patting her cheek. But then she sat down again, lost. My father put his arm around her.

That night I stayed after the others left. Later in the evening we sat in the spare living room, the curtains drawn, the air conditioner running full blast. I was chilled, but Grandma suffered terribly from the August heat.

She laid her head back on the rocker, her eyes closed. We

sat listening to the air blowing. Then she leaned forward and pulled a dog-eared sign language book from the drawer in a table between the two easy chairs. It was hidden underneath the photo albums.

"I've always been disappointed that I didn't learn the sign language more," she said. "I tried, but it just didn't come naturally. . . . See? I can do them to myself," and she proceeded to sign "apple," "cake," "rum," and other words from diagrams, paging through the book, finding signs she liked, trapping the page with her elbow while she did the gesture, her palms toward her face as if she were signing only to herself.

"How come you don't do this when Mom and Dad are here?" I asked.

"Oh, when I get around the kids, I just can't seem to remember. Your mother and dad go so fast. They wouldn't want to wait for me."

It was dizzying, entering that room at the funeral home, with green and white mausoleums on the wallpaper, rosy swirls in the carpet, and gold brocade on the chairs. I walked to the front, where Grandpa's polished oak casket was nestled in a profusion of yellow spider mums and orange gladiolus. My grandfather had loved fishing. Most of the baskets were adorned with miniature rods and reels. Mom stood next to me, her arms on mine. I could hardly look into the open casket.

Suddenly, my mother burst into sobs, soul-wrenching, piercing cries such as I'd never heard before in my life. She doubled over, her body trembling.

"Should we get a doctor? A sedative?" Grandma asked, scared, blinking. "What should we do?"

The funeral director rushed in. He was a thin man with an oversized black mustache. "Would she like some water? Some coffee?"

Jan and I led Mom to a chair. Her face was buried in her hands and she was shaking her head back and forth.

"Can't we do something?" Grandma asked.

"It's all right," I told her. "She's okay."

Grandma, hands clenched, not quite sure what to do, walked back over to my grandfather. Jan looked up at me. We both had our hands on Mom's heaving back, trying to soothe her. "Grandma doesn't know that Mom just can't hear herself," Jan whispered. "She hasn't lived with her in a long time. Mom doesn't know how loud she is."

My mother had just lost her father. No relationship between parent and child is simple. And no matter how eloquent we are, grief cannot be translated into words. In her mind, the images of their lives were replaying, speeded up and slowed down. Every moment of their lives. The characteristic gestures and expressions. The things she'd depended upon seeing. She was in agony.

Later on that afternoon, Mom said she was worried about Grandma. "I wish we lived closer so I could see her more," Mom signed. "I don't want her to be lonely."

Then Mom asked if she'd been too noisy, a sign made with the hand shaking at the ear.

"No, don't worry about it," Jan signed to her. "Please don't. It was normal."

Callers began arriving at the funeral home even before the appointed two o'clock visiting hour, and it was a very long six hours until the last of them left. Despite the fact that I had spent all those summers with Uncle Bill hanging around the funeral home, I'd never had to stand watch before. Jan and I took turns interpreting for Mom and Dad. Every once in a while we'd walk downstairs to a private lounge, to sit and rest our eyes and arms. Grandma never left the room the entire day, and sat down only once, for a few seconds in the chair next to Grandpa's casket.

Mom knew most of the people from when she was a little girl growing up in Greencastle and Fillmore. One woman reminded Mom of how she used to play with her daughter, how she'd pull her little red wagon in front of their house every afternoon. Mom and Dad lip-read the people and looked over at Jan or me when there was something they missed in a sentence. Mom's memory was extraordinary. Watching all those mouths was exhausting because some of the people talked nonstop, but she could honestly recall every name and incident.

At one point an obviously demented woman got hold of my mother, grabbing her hands, thrusting her face into Mom's, talking a blue streak. Mom shot me a look that asked what the woman was saying. I interpreted, but Mom looked confused.

"She's not making any sense," I said. "Crazy," I signed, my hand twisting an imaginary bolt back and forth at the side of my head as if a screw were loose.

Friends of Mom's and Dad's arrived in the early evening.

"That was nice of their deaf friends to come," Grandma said. "I'm glad they had someone to talk to."

The next morning at the funeral, it was decided I would interpret. It seemed fair. I got home the least frequently and Jan and Kay had recently signed other difficult situations. Before the sermon began, I tried positioning myself so that my mother and father could see the preacher and the casket as well as look at me.

"She shouldn't stand up," my mother said to my father. "Everyone will stare."

"But all these people know we're deaf," he signed patiently. "It's the way it must be done. We need to be able to see."

My mother calmed down. As I signed the service, I tried not to think of the words on my hands. I didn't want to get emotional as I conveyed my grandfather's funeral sermon to my mother and father.

Standing there, taking the invisible words from the air and placing them for my mother and father to see, I searched my mother's face for echoes of the face of the man lying in the box a few feet behind me.

"Ashes to ashes. Dust to dust. We come here to bury Chester Cooper Wells." I signed "bury" with two open hands lowering him straight down into the earth—but gently, gingerly. I didn't want to hurt my mother any more than I had to.

I did the mental counting up all people must do at funerals. I thought about my relatives who had died: H.T., Nellie May, Aunt Margaret. I thought about the ones who were alive: my parents, sisters, aunts and uncles, my grandmother, friends. I could hear my Grandpa Wells's voice and his self-doubts. "I don't know if I done right. I don't know if I done all I should," he'd say, sitting on the back step, resting his elbows on his knees. He'd been cutting up an apple. I was little; I wasn't quite sure I knew what he was talking about. Then he'd try to dismiss the heavy mood, offering me an apple section on the tip of a pocketknife. What do you do about the past? He'd missed so much of my mother's growing up, of her conversation, of her. But he'd also cared for her and he'd tried to tell her that once. Even though life got in the way.

I wanted him alive. For my mother. For my grandmother. I wanted to get to know him better as an adult, even though if he hadn't been my grandfather I doubt we would ever have been friends. I doubt our lives would have crossed. Strangely, I was most grateful for the past few years, the ones after the Christmas of the missed connections. And most of those years I'd spent angry at him and hurt by him—for his trying to tell his daughter he loved her. And failing. I was relieved that when he died I was no longer angry or hurt, that he'd lived long enough for me to realize that people are not bad or good, but that all things and all people are complicated layers of good and bad and confused. I'd wanted to explain to him all I'd learned about what deafness was and what it did. But as the minister's words tried to establish a peace for my grandfather, I made my own peace.

There was no use brooding about it any longer. I'd seen plenty of families where there was more communication and less love.

After the trip to the cemetery, we headed back to my grandmother's house. Neighbors kept stopping by with casseroles and cakes.

"It was as if Grandpa planned it this way. He wouldn't want to get in the way of Jan's wedding," Grandma said. "He wouldn't want to upset anyone else's plans. No, he never would."

Grandma was sitting in the backyard, holding the baby, her great-grandchild. She looked down at him. "I've had some real shocks in my life. Doris's being deaf. Grandpa's dying so quick. My brother Lee, sister Pauline, brother Tim, dying so young, all of them. Yes, I've really had some terrible shocks."

I was signing this to my mother. She didn't say anything. I wished Grandma hadn't made Mom the first of the list.

"But I lived through them," she said, "and I'll live through this, I suppose."

As we drove away, Grandma stood alone on the porch, waving.

Epilogue

September 1983

The funeral was Monday; Jan's wedding was Saturday.

It was hard to ignore our sadness. We threw ourselves into the preparations and tried not to think about what had gone on before. Jan and her fiancé, Brian, were superbly organized, but there were still plenty of errands to do that week. Mom and I made rice packets out of white net and pink crinkle ribbon. Dad picked up the champagne fountain and took it to the reception hall. We had last-minute fittings for the bridesmaids' dresses, which Jan had sewn. The four of us looked like English milkmaids with our white eyelet frocks, pink sashes, and white stockings.

There was a rehearsal and a dinner Friday night. We'd hired an interpreter to stand at the altar for Friday and the Saturday ceremony, since both Kay and I were going to be bridesmaids. The minister began the rehearsal but had to stop in order to find a box for the interpreter to stand on; she was too short to be seen over the podium.

By rights the rehearsal dinner is an uproarious occasion. This one was subdued and dignified. Brian's father and the best man made solemn toasts. Dad began his with a small, simple joke. I voiced what he said.

"I must admit that when Brian first wanted to marry Jan," he signed, "I was suspicious. I thought perhaps he just wanted to borrow my luggage." (Brian had used my parents' suitcases, the chestnut-brown ones they'd bought for their honeymoon, on many occasions.)

Dad put down his champagne glass so that he could sign more eloquently, using both hands.

"I am kidding you. I am proud to have Brian join our family. I know I am not losing my daughter. I love her too much. Brian will be a good part of our family. I know Jan will be happy in her marriage and that Brian will be happy in his marriage. To a long, wonderful life."

We raised our glasses and drank.

My interpretation was inadequate. My father's signing was graceful and expansive. It had the beauty of a conductor leading a symphony orchestra. There was nothing clichéd in what he signed. No translation could have been as expressive or as moving as the way he drew his hands through the air. They were gestures made in public, but the meaning was private and loving.

The morning of the wedding was exactly as the morning of any wedding must be: a flurry of preparations. Mom had laid out her clothes and Dad's. She wanted to press my dress, but I wouldn't let her. Dad couldn't find his studs. I went wildly searching for them. His shirt needed ironing and the sleeves were too long.

We got to the church in plenty of time. Hiding in the choir loft, Jan, Kay, and I watched the ushers seating people. We exchanged sisterly whispers, the three of us surprised at some who showed up.

Jan wore a delicate turn-of-the-century wedding dress of silk and inset lace, with a high Victorian collar. Dad walked her down the aisle as if she really were the china doll Grandma Wells always said she was. Standing before the minister, Dad watched the interpreter intently and when the minister asked: "Who gives this woman to be wed?" Dad answered aloud, quietly, slowly, but clearly, signing as he spoke: "Her mother and I do."

The reception was held in a large, greenery-filled clubhouse, and the party quickly spread to several rooms. Every once in a while I'd look over, to see my grandmother alone, fighting back her tears. She didn't want to grieve during her grand-daughter's wedding.

A natural division occurred at the party. The deaf people sat in one room, catching up on the gossip, enjoying themselves, while the hearing people stayed in the room with the buffet and the band.

Dad and Mom were perfect hosts, never hovering, but congenial. I adored it when Mom was all dressed up and excited. There was an aura about her, a beauty that transcended her features. I thought back to the letter she'd written me after Brian formally, in sign, had asked my father for Jan's hand in marriage. "We are pink tickled," she wrote. "Brian is so fine young man." And here today Mom looked lovely in her flowing cream-colored dress, her once red hair now a lustrous gold. Dad still had his boyish cowlick. A grin emphasized the cleft in his chin. Somehow it didn't matter that his rented tux didn't quite fit.

Every once in a while Mom or Dad would walk into the main reception room to see how the party was going. Mom danced with Dad and then she danced with Brian and Brian's father, feeling the heavy bass beat through the floorboards. Dad danced with Jan, then Kay, then me, and was happy to let us lead.

My parents shuttled back and forth between the two parties. The interpreter had left quite a while ago, but in situations like these, interpreters really aren't needed. There was something in my parents' smiles, in the way they were holding themselves, that made everything seem fine. Dad was proud of his family, proud of the fine wedding and reception, delighted that so many of his relatives and friends had come. The communication was basic. People walked up to Mom and Dad, beaming, pumping their hands, pointing to Jan and Brian and holding up a rounded thumb and forefinger—a universal "OK" sign.

Dad came in and danced a couple of big-band tunes with me. I introduced him to twirls and swoops and he was enjoying himself immensely. "I follow you," he signed. I asked him if he knew how to dip.

"What's that?"

Signing and doing a slight gesture, I explained that at the end of the song he would tilt me backward.

"I follow you," he signed again in his most courtly manner, a curled right hand acting as if it were following a curled left. It turned out to be a very successful dip.

"I was afraid I'd drop you," he said, kissing me on the cheek.

After our dance I made the rounds in both rooms, chatting in sign with friends of Mom's and Dad's I'd known since childhood. Back in the main room near the band, I looked up to see Mom and Dad standing at the door, surveying the scene. He had his arm around her. She was leaning against him. I thought about that shy young couple in the Great Smoky Mountains thirty years earlier. And all the disappointments and frustrations they'd had before and after. Yet as I'd watched other marriages crumble or freeze into icy acquaintanceships, theirs was different. They were very much together.

In New York I'd just met a wonderful man and I was falling in love. Helplessly, happily. This was different from my other loves. It was uncomplicated and sweet. And I could only hope that it would have a future as long and that the relationship would be as strong as my parents'. I'd started this one clean and fresh. I'd told him about my parents, not in that gasping, embittered confessional tone but through stories, the stories of my growing up and Mom's and Dad's growing up. For me, the past had finally emerged from being a horrible, dark secret to being an unusual family's history. I was altering my perception to make life happier and easier to live, and that change was working well. I was even looking forward to bringing my beau back home for Christmas.

It was a fine celebration.

Dad led Mom to the dance floor. They stole a couple of glances at the dancers around them to make sure the beat hadn't changed—or the music stopped. After a few cautious steps they danced to whatever beat they felt like, wrapped in each other's arms. Dad twirled Mom around and held her as they dipped.

And then he looked up at me and winked.